1+X 职业技术·职业资格培训教材

计算机操作员

文字录入

主　编　成　戈

编　者　徐春凌　陆　辉　许　玫　冯海英　陈玉红

主　审　陈丽娟　张士忠

中国劳动社会保障出版社

图书在版编目(CIP)数据

计算机操作员：四级．文字录入/上海市职业技能鉴定中心组织编写．—3 版．—北京：中国劳动社会保障出版社，2014

1+X 职业技术·职业资格培训教材

ISBN 978-7-5167-0923-8

Ⅰ.①计… Ⅱ.①上… Ⅲ.①电子计算机-技术培训-教材②文字处理-技术培训-教材 Ⅳ.①TP3

中国版本图书馆 CIP 数据核字(2014)第 092847 号

中国劳动社会保障出版社出版发行

(北京市惠新东街 1 号　邮政编码：100029)

*

三河市潮河印业有限公司印刷装订　　新华书店经销

787 毫米×1092 毫米　16 开本　10.75 印张　204 千字

2014 年 5 月第 3 版　　2019 年 6 月第 3 次印刷

定价：25.00 元

读者服务部电话：(010)64929211/84209101/64921644

营销中心电话：(010)64962347

出版社网址：http://www.class.com.cn

内容简介

本教材由人力资源和社会保障部教材办公室、中国就业培训技术指导中心上海分中心、上海市职业技能鉴定中心依据上海 1+X 计算机操作员（四级）职业技能鉴定细目组织编写。教材从强化培养操作技能，掌握实用技术的角度出发，较好地体现了当前最新的实用知识与操作技术，对于提高计算机操作员基本素质，掌握计算机操作员（四级）的核心知识与技能有直接的帮助和指导作用。

根据本职业的工作特点，以能力培养为根本出发点，全书分为 3 个项目共计 18 个活动，内容涵盖英文录入、中文录入和中英文混合录入。每个活动都由一段课前小短文引出，由“摩拳擦掌”“师傅领进门”“修行靠个人”“灵丹妙药”和“过关斩将”5 个部分组成，其中，“摩拳擦掌”介绍学习本活动前需了解的基础知识；“师傅领进门”涵盖了本活动的主体内容，是该学习部分的核心；“修行靠个人”给出若干分组练习，使学员能够在重点难点上进行专项的强化训练；“灵丹妙药”则介绍了一些避免差错和提高技巧的方法；最后设置的“过关斩将”环节则能帮助学员测试自己的学习成果。此外，每个项目结尾的“藏经阁”和“打擂台”更能帮助学员对本项目的学习内容进行总结归纳。

本教材可作为计算机操作员（四级）职业技能培训与鉴定考核教材，也可供全国中、高等院校计算机操作相关专业师生参考使用，以及本职业从业人员培训使用。

改 版 说 明

计算机操作员职业以个人计算机及相关外部设备的操作为常规技术和工作技能，是国家计算机高新技术各专业模块的基础。随着信息技术的不断发展，计算机操作员的职业技能要求有了新的变化。2014 年上海市职业技能鉴定中心组织有关方面的专家和技术人员，对计算机操作员职业进行了提升，计算机操作员分为五级、四级两个等级，其中四级又细分为系统管理、办公软件应用、文字录入三个方向。

为了更好地为广大学员参加培训和从业人员提升技能服务，人力资源和社会保障部教材办公室、中国就业培训技术指导中心上海分中心与上海市职业技能鉴定中心组织相关方面的专家和技术人员，依据新版计算机操作员（四级）文字录入职业技能鉴定细目对教材进行了改版。新版教材改变了原先“学习单元”和“节”的传统编写模式，采取“项目”和“活动”的形式，通过虚构的学员“小文”和“小璐”从零开始学习计算机文字录入的过程，由简到繁、由浅入深，寓教于乐，让学员在故事中逐步学习并领悟文字录入的原则、方法和技巧。此外，新版教材为了跟上计算机技术的发展和职业鉴定考试的提升要求，对原教材在内容上也进行了重大调整，不再对 Windows 操作系统和 Word，Excel，Outlook 等应用软件进行介绍，而把主要的目光聚焦在“文字录入”尤其是“五笔字型输入法”的各项训练活动上，更加注重技能培养，更加切合“文字录入”职业的需要。

前　　言

职业培训制度的积极推进，尤其是职业资格证书制度的推行，为广大劳动者系统地学习相关职业的知识和技能，提高就业能力、工作能力和职业转换能力提供了可能，同时也为企业选择适应生产需要的合格劳动者提供了依据。

随着我国科学技术的飞速发展和产业结构的不断调整，各种新兴职业应运而生，传统职业中也愈来愈多、愈来愈快地融进了各种新知识、新技术和新工艺。因此，加快培养合格的、适应现代化建设要求的高技能人才就显得尤为迫切。近年来，上海市在加快高技能人才建设方面进行了有益的探索，积累了丰富而宝贵的经验。为优化人力资源结构，加快高技能人才队伍建设，上海市人力资源和社会保障局在提升职业标准、完善技能鉴定方面做了积极的探索和尝试，推出了1+X培训与鉴定模式。1+X中的1代表国家职业标准，X是为适应经济发展的需要，对职业的部分知识和技能要求进行的扩充和更新。随着经济发展和技术进步，X将不断被赋予新的内涵，不断得到深化和提升。

上海市1+X培训与鉴定模式，得到了国家人力资源和社会保障部的支持和肯定。为配合1+X培训与鉴定的需要，人力资源和社会保障部教材办公室、中国就业培训技术指导中心上海分中心、上海市职业技能鉴定中心联合组织有关方面的专家、技术人员共同编写了职业技术·职业资格培训系列教材。

职业技术·职业资格培训教材严格按照1+X鉴定考核细目进行编写，教材内容充分反映了当前从事职业活动所需要的核心知识与技能，较好地体现了适用性、先进性与前瞻性。聘请编写1+X鉴定考核细目的专家，以及相关行业的专家参与教材的编审工作，保证了教材内容的科学性及与鉴定考

核细目以及题库的紧密衔接。

职业技术·职业资格培训教材突出了适应职业技能培训的特色，使读者通过学习与培训，不仅有助于通过鉴定考核，而且能够有针对性地进行系统学习，真正掌握本职业的核心技术与操作技能，从而实现从懂得了什么到会做什么的飞跃。

职业技术·职业资格培训教材立足于国家职业标准，也可为全国其他省市开展新职业、新技术职业培训和鉴定考核，以及高技能人才培养提供借鉴或参考。

新教材的编写是一项探索性工作，由于时间紧迫，不足之处在所难免，欢迎各使用单位及个人对教材提出宝贵意见和建议，以便教材修订时补充更正。

人力资源和社会保障部教材办公室
中国就业培训技术指导中心上海分中心
上 海 市 职 业 技 能 鉴 定 中 心

目 录

CONTENTS

风云榜

项目一 英文录入

4 活动 1 稳扎稳打——键位练习

20 活动 2 百发百中——纠正指法

30 活动 3 运指如飞——文章练习

43 藏经阁

43 打擂台

项目二 中文录入

46 活动 1 改头换面——录入中文

50 活动 2 步步为营——录入字根

61 活动 3 名不虚传——录入键名汉字

64 活动 4 一笔一画——录入成字字根

70 活动 5 庖丁解牛——键外字拆分

80 活动 6 一锤定音——末笔识别码

88 活动 7 精益求精——拆字技巧

94 活动 8 一马当先——录入一级简码

97 活动 9 合二为一——录入二级简码

105 活动 10 熟能生巧——录入常用字

114 活动 11 事半功倍——录入词组

CONTENTS

123 活动 12 融会贯通——录入文章
130 藏经阁
130 打擂台

项目三 中英文混合录入

134 活动 1 随机应变——状态切换
141 活动 2 无所不知——特殊符号录入
149 活动 3 大功告成——文章练习
155 藏经阁
155 打擂台

附录一 成字字根编码表 / 157

附录二 末笔识别码编码表 / 159

附录三 常用千字 / 163

FENGYUNBANG

风云榜

最新报道：2013年7月12日至19日，在比利时根特举行了第49届“国际速录大赛”，世界各国的打字高手们齐聚一堂，切磋技艺，中国国家队也参加了这次比赛，并取得了辉煌的战绩……

“小璐，小璐，快来看呀！今年的国际大赛有消息了！”小文一边翻看着网上的新闻，一边兴奋地叫道。

“终于有结果了！今年我们国家队又取得了哪几项冠军呀？”小璐急匆匆地跑过来，“我们快去告诉师傅吧。”

……

书房里，师傅正在闭目养神，一旁的书架上赫然摆放着很多金光灿灿的奖牌和奖杯。听见门外徒弟们吵吵嚷嚷的声音，他微笑着打开房门。

“你们也看到今年‘国际速录大赛’的最新消息了吧。”师傅不紧不慢地说，“这可是国际顶级的大赛，很多‘吉尼斯世界纪录’都是在这项赛事中诞生的。我们国家虽然这几年才开始参加这项比赛，但是成绩却一鸣惊人，曾经在第47届比赛中拿到团体总分第一名呢！”

“师傅，师傅，你快给我们讲讲吧！”

“既然你们感兴趣，我就给你们开开眼界吧！”师傅打开书桌上的笔记本电脑，世界各地的打字高手和赛事资料应有尽有……

NO.1 “世界打字女王”“吉尼斯世界纪录”保持者马特什科娃。在2001年德国汉诺威市举行的第43届“国际速录大赛”中，这位来自捷克的女秘书在30分钟内共敲下24 224个键，其中包括英文字母、数字和各种符号，平均每分钟敲键807个，而且字字准确无误，打破了之前保持16年之久的原“吉尼斯世界纪录”。

NO.2 原“吉尼斯世界纪录”保持者布莱克本。在1985年的“国际速录大赛”中，布莱克本在50分钟内准确地敲下37 500个键，从而以平均每分钟敲键750个的成绩创下当时的“吉尼斯世界纪录”，并保持这一纪录长达16年。

NO.3 2009年，第47届“国际速录大赛”在中国北京成功举办，这是国际速联自成立以来首次在亚洲举办大赛。中国代表团在10个比赛项目中获得金牌16枚、银牌12枚、铜牌13枚、位列团体总分第一名。

NO.4 2011年，中国速录国家队17名选手出征在法国巴黎举办的第48届“国际速录大赛”，共获得7项冠军、3项亚军和8项季军的优异成绩。

……

看着两个目瞪口呆的徒弟，师傅好像早在预料之中，“好了，下面该轮到我们了，只要你们认真学习，坚持练习，想要超过师傅，甚至是向这些世界级的高手挑战，都是指日可待的事，加油吧！”

IANGMUYI

项目一 英文录入

活动1 稳扎稳打——键位练习

这天一大早，小文和小璐就等在师傅的书房里准备开始学艺了。

“都准备好了吗？所谓磨刀不误砍柴工，要想练好打字这门功夫，首先必须把基本功——‘指法’修炼扎实，我们的目标就只有一个：实现‘盲打’！”师傅说。

“盲打？”小璐不解地问道，“什么是盲打呢？难道看不见还能打字吗？”

“盲打，就是眼睛不看键盘，凭手指的感觉去正确击键，是实现高速、准确录入的必备技能。”师傅解释道。

“那不是很神奇嘛！不用看就能想哪打哪，就像是武林高手！”小文惊奇地说，“不过，这一定会很难吧？”

“其实盲打一点儿也不难！只要掌握正确的指法，每个人都能很快做到盲打。而指法和游泳、骑自行车一样，都属于需要在短期内克服困难才能掌握的，但也是只要稍加努力就一定能够掌握的基本技能，并且掌握之后就终生不会忘记和退化。”师傅目光如炬地紧盯着两个孩子，“你们有信心克服困难，练成这项终身受益的技能吗？”

小文和小璐对视了一眼，不约而同地大声回答：“我们一定行！”

摩拳擦掌

在开始学习指法之前，首先认识一下我们手中的“武器”——键盘。我们常用的键盘被称作 QWERTY 键盘，它是用键盘字母键区第一行的前 6 个字母按键来命名的。由于生产厂商和品牌型号的不同，键盘也会有一些差别，但功能基本一样，都包含 4 个主要区域：主键盘区、功能键区、控制键区和数字键区，如图 1—1—1 所示。

图 1—1—1 键盘主要区域分布

其中，主键盘区是键盘最重要的组成部分，也是进行文字录入主要使用的区域。所以再把主键盘区细分成字母、数字、标点符号和功能键等几个区域，如图 1—1—2

所示。熟悉了键盘的所有区域分布以及每个区键位的排列规律，就像是在心里有了一张地图，能够在文字录入时为你的手指指引方向。

图 1—1—2 主键盘区

需要特别注意的是，字母区字键的排列位置并不是按 ABCDEF……的英文字母顺序排列的，而是按照各英文字母在文章中出现频率（见表 1—1—1）的高低来排列的。

表 1—1—1 **英文字母使用频率表**

字母	频率	字母	频率	字母	频率
A	8.19%	J	0.14%	S	6.36%
B	1.47%	K	0.41%	T	9.41%
C	3.83%	L	3.77%	U	2.58%
D	3.91%	M	3.34%	V	1.09%
E	12.25%	N	7.06%	W	1.59%
F	2.26%	O	7.26%	X	0.21%
G	1.71%	P	2.89%	Y	1.58%
H	4.57%	Q	0.09%	Z	0.08%
I	7.10%	R	6.85%		

如果我们把 26 个字母按照出现频率的高低分为三类：高频字母 A，E，I，N，O，R，S，T（深灰色）；中频字母 C，D，F，H，L，M，P，U（浅灰色）；低频字母 B，G，J，K，Q，V，W，X，Y，Z（白色字）。通过字母使用频率和键盘位置的对照比较，大家可能会惊奇地发现：使用频率高的字母和使用频率低的字母全部被分散开来，散落在键盘各处，如图 1—1—3 所示。

为什么会这样排列字母呢？

答案是：QWERTY 键盘就是为了降低打字速度而设计的！

QWERTY 键盘最初是为打字机而发明的，一开始打字机的键盘是按照字母顺序排列的，但如果打字速度过快，某些键的组合很容易出现卡键问题。于是“打字机之

图 1—1—3 键盘字键使用频率

父”——美国人克里斯托夫·拉森·授斯发明了 QWERTY 键盘布局，他将最常用的几个字母安置在相反方向，最大限度地放慢敲键速度以避免卡键。授斯在 1868 年为该键盘申请专利，1873 年使用此布局的第一台商用打字机成功投放市场，并一直沿用至今，成为了电脑的基本输入设备——键盘的通用标准。

虽然 QWERTY 键盘的安排方式非常没效率，但它是标准。理解它的设计思路，从而克服这些先天的困难，对于我们练习打字是非常有益的，比如：

● 使用 QWERTY 键盘，右手负担了 43%的工作，而左手却负担了 57%的工作。意味着我们练习打字时一定要加强左手练习。

● 使用 QWERTY 键盘，需要频频使用小指和无名指，而一般人的两手小指及左手无名指力量最小，平时必须加强这三根手指的击键力量练习。

● 使用 QWERTY 键盘，排在中列的字母，其使用率仅占整个打字工作的 30%左右。因此，在打字过程中时常要上上下下移动指头，在移动中还要准确盲打击键，这就必须依赖科学的指法。

师傅领进门

一、基准键

在 QWERTY 键盘上，排在中行的左右各 4 个键被称作基准键，也称导键。分别是左侧的 4 个键（【A】【S】【D】【F】）和右侧的 4 个键（【J】【K】【L】【;】），如图 1—1—4 所示。

二、范围键

与 8 个基准键相对应，把主键盘区分成 8 个区域范围，每个范围内的其他字键称为范围键，如图 1—1—5 所示。

三、标准指法一

静止状态，双手从食指到小指的四根手指始终悬停在基准键上：即让你的左手食

图 1—1—4　键盘基准键

图 1—1—5　键盘范围键

指放在【F】上（【F】键上有一个小突起，我们通常称为盲打坐标），右手食指放在【J】上（【J】键上也有一个盲打坐标），然后将四指并列对齐分别放在相邻的键位上，双手大拇指都摆放在空格键上，对应关系如图 1—1—6 所示。

四、标准指法二

手指与键盘位置的“双垂直”原则：即手指的第一指节与键面尽量垂直，双手手指与导键的位置要接近垂直，不可成“八”字形。否则，击键时容易打在两个字键的中间位置而造成连键，也不好发力，如图 1—1—7 所示。

五、标准指法三

打字过程中的“倾斜”移动原则：即每根手指负责击打其悬停的基准键对应区域的所有范围键，在击打范围键时，无论左手还是右手，都要遵从“左高右低”的方式上下倾斜移动。每根手指需要管辖的范围键必须牢记，位置如图 1—1—8 所示。

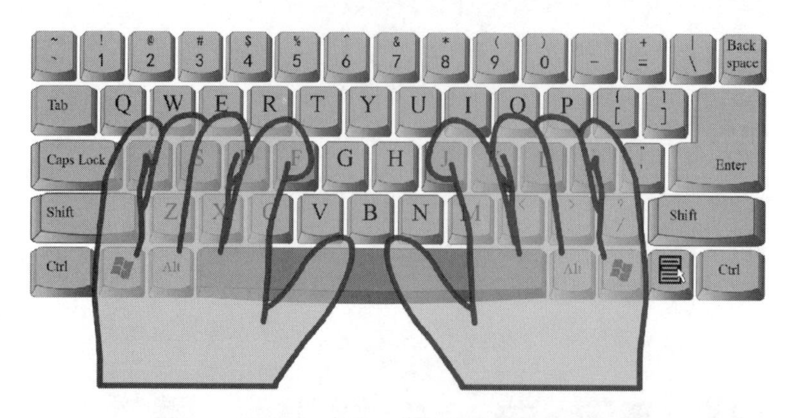

图 1—1—6　双手静止悬停状态

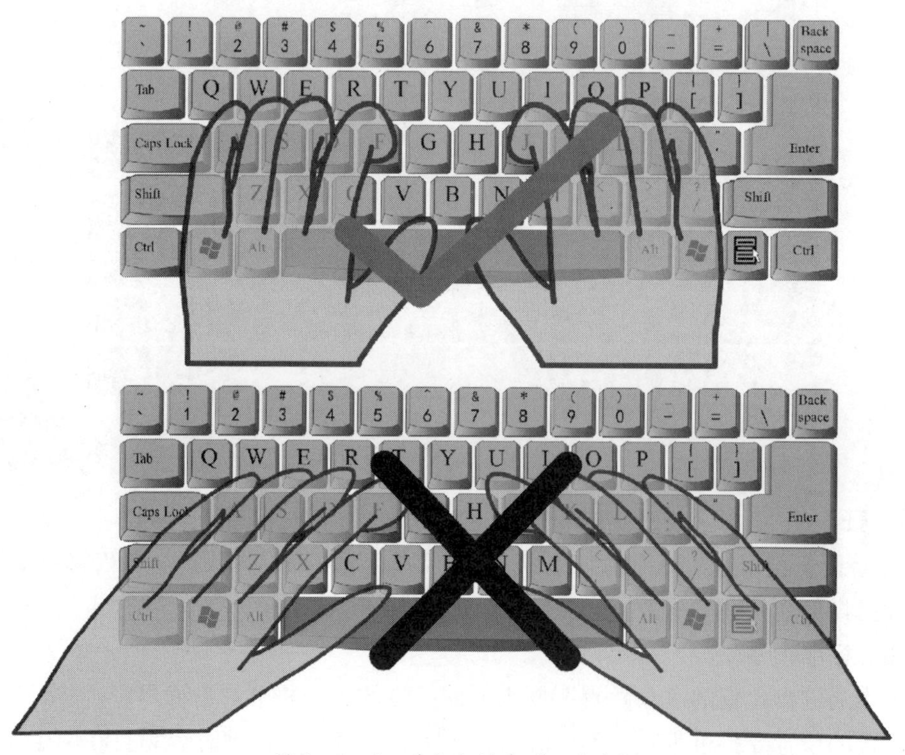

图 1—1—7　手指与键盘“双垂直”

六、标准指法四

打字过程中手指的“独立”移动原则：即一根手指在击键过程中独立运动，其余不击键的手指不要跟着一起离开基准键位，也不要翘起，如图 1—1—9 所示。

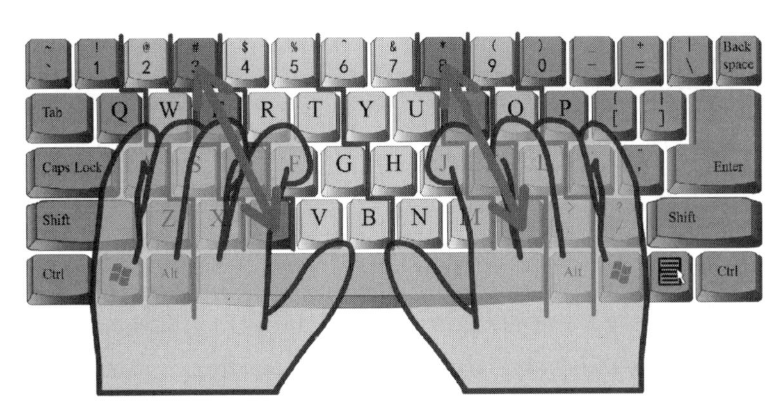

图 1—1—8 倾斜移动轨迹

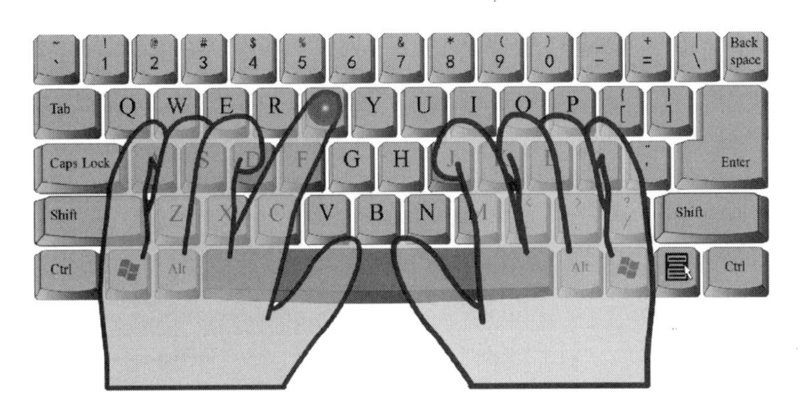

图 1—1—9 独立移动轨迹

七、标准指法五

打字过程中的“手势还原”原则：即每一根手指在移动击键的任务完成后，一定要习惯地回到基准键位置，就是恢复“静止状态”。

八、常用功能键指法：空格键、回车键【Enter】、回退键【Backspace】

● 空格键：空格键是文字录入过程中使用频率最高的按键，可以用两只手的大拇指敲击，左、右手随意。

● 回车键【Enter】：回车键也是使用频率较高的按键，由右手的小拇指进行敲击。

● 回退键【Backspace】：回退键也称删除键，是在文录过程中需要修改已录入内容时使用的按键，依据录入员输入的准确性高低来决定它的使用频率，对于初学者来说使用频率肯定偏高，也由右手的小拇指进行敲击。

从以上可以看出，右手小拇指在文录过程中负担很重，平时可以用指尖连续敲击桌面，手指交错上下起伏运动，或反复攥拳、伸掌等方式加强力量和反应练习。

修行靠个人

要求：反复练习以下每部分的输入内容，直到该部分的目标达成以后，才能进入下一部分的输入练习！

一、基准键输入练习

1. 输入以下字母（左右手对应手指练习）

目标：录入速度达到120字符/分钟，即在2分钟以内完成录入。

jjff jjff ffjj ffjj fjfj fjfj jfjf jfjf
ddkk ddkk kkdd kkdd dkdk dkdk kdkd kdkd
ssll ssll llss llss lsls lsls slsl slsl
;;aa ;;aa aa;; aa;; a;a; a;a; ;a;a ;a;a
fjkd fjkd lsa; lsa; slfj slfj ;akd ;akd
jf;a jf;a kdls kdls a;dk a;dk slfj slfj

2. 输入以下字母（强化练习）

目标：录入速度达到90字符/分钟，即在5分钟以内完成录入。

asdf ;lkj asdf ;lkj asdf ;lkj
fjds klsa fjds klsa fjds klsa
kfjd als; kfjd als; kfjd als;
sajd flk; sajd flk; sajd flk;
lkds asl; lkds asl; lkds asl;
akd; jfls akd; jfls akd; jfls
;ksf adjl ;ksf adjl ;ksf adjl
j;al fskd j;al fskd j;al fskd
lkjd fas; lkjd fas; lkjd fas;
ksfa dlj; ksfa dlj; ksfa dlj;
fsad jk;l fsad jk;l fsad jk;l
;fld ajsk ;fld ajsk ;fld ajsk
lsf; dkja lsf; dkja lsf; dkja
afj; lkds afj; lkds afj; lkds
adlj ;kfs adlj ;kfs adlj ;kfs

3. 输入以下单词（综合练习）

目标：录入速度达到90字符/分钟，即在3分钟以内完成录入。

ass	ass	ass	ass	ass
sad	sad	sad	sad	sad
flask	flask	flask	flask	flask
lad	lad	lad	lad	lad
lass	lass	lass	lass	lass
fad	fad	fad	fad	fad
all	all	all	all	all
salad	salad	salad	salad	salad
add	add	add	add	add
ask	ask	ask	ask	ask
fall	fall	fall	fall	fall
dad	dad	dad	dad	dad

二、范围键【G】【H】【R】【T】【Y】【U】输入练习

1. 输入以下字母（左右手对应手指练习）

目标：录入速度达到120字符/分钟，即在2分钟以内完成录入。

gghh	gghh	hhgg	hhgg	ghgh	ghgh	hghg	hghg
ttyy	ttyy	yytt	yytt	tyty	tyty	ytyt	ytyt
uurr	uurr	rruu	rruu	ruru	ruru	urur	urur
ghur	ghur	ythg	ythg	ruty	ruty	ghty	ghty
ghfj	ghfj	ytjf	ytjf	rufj	rufj	hgjf	hgjf
rua;	rua;	ytls	ytls	ghdk	ghdk	kdyt	kdyt

2. 输入以下字母（强化练习）

目标：录入速度达到90字符/分钟，即在5分钟以内完成录入。

dsft	dfjt	dytf	ydug	hfuk	dfjr
frhu	fyrh	duty	tl;s	rytj	tyug
rjgu	jast	furk	frau	fhtj	kdsr
ufgt	fru;	fyth	kdur	hjrs	hyrl
ujfy	jutf	jyuh	ftkr	fytr	klyu
hrtk	atrk	dsyt	kysr	fyrd	slyg
adug	hrlk	;lrg	utha	frgs	rsda
hudr	ksyg	jrha	tfhg	liud	ksuh
fthy	jrhu	hsys	lkur	jrya	lifa

dist	lrkt	sita	luth	ditf	guyr
gtrh	hyug	jidt	ks;t	ufgh	thlk
h;ly	gats	guls	lk;r	rthu	hfjt
jsug	sthg	lr;t	ausy	sulr	;tau
jghf	uryt	furj	hygt	ljyu	sfjr
jthf	gurf	jtyh	futh	lsur	fgys

3. 输入以下单词（综合练习）

目标：录入速度达到 90 字符/分钟，即在 3 分钟以内完成录入。

glad	glad	glass	glass	had	had
flag	flag	half	half	august	august
hall	hall	salt	salt	shut	shut
star	star	stay	stay	dusk	dusk
that	that	rust	rust	gay	gay
fast	fast	duty	duty	full	full
lady	lady	far	far	hurry	hurry
haul	haul	hart	hart	flat	flat
has	has	hard	hard	taught	taught

三、范围键【Q】【W】【E】【I】【O】【P】输入练习

1. 输入以下字母（左右手对应手指练习）

目标：录入速度达到 120 字符/分钟，即在 2 分钟以内完成录入。

qqpp	qqpp	ppqq	ppqq	qpqp	qpqp	pqpq	pqpq
wwoo	wwoo	ooww	ooww	owow	owow	wowo	wowo
eeii	eeii	iiee	iiee	eiei	eiei	ieie	ieie
qpwo	qpwo	ieow	ieow	pqie	pqie	qpty	qpty
owur	owur	iehg	iehg	qpa;	qpa;	wosl	wosl
iekd	iekd	eifj	eifj	wogh	wogh	pqhg	pqhg

2. 输入以下字母（强化练习）

目标：录入速度达到 90 字符/分钟，即在 5 分钟以内完成录入。

qaws	wsed	edik	ikol	olp;
p;qa	qeip	pwqo	qwef	iopj
ropw	qide	lodq	wije	olkp
ieop	ikpq	wios	paos	wlqe

oise	urew	osle	;aew	yuio
trew	piuw	oskj	pwjd	hijo
sewq	hgie	orek	peoq	iods
qwjl	hurw	olse	wieq	wpot
riqy	thew	hrpq	pwit	wpoq
iwje	elsi	sile	iske	iasw
pwer	utyi	ieur	hiew	poej
ewlk	peih	sfew	arwq	srwp
rtew	wqor	hrei	oeiw	owle
ueoi	opih	ewfq	ohge	oupq
wrei	qepg	grew	ioup	egfi
pioe	siew	oerw	peor	irks
oewj	;aoe	roq;	ojwf	pqfj
oejf	wfoj	pqwe	owqf	piej

3. 输入以下单词（综合练习）

目标：录入速度达到 90 字符/分钟，即在 3 分钟以内完成录入。

hold	hold	pass	pass	quart	quart
told	told	will	will	protest	protest
with	with	worker	worker	operate	operate
equal	equal	world	world	side	side
dead	daed	allow	allow	push	push
wall	wall	people	people	deft	deft
okay	okay	warp	warp	were	were
work	work	patk	park	look	look
puddle	puddle	soap	soap	write	write

四、范围键【V】【B】【N】【M】输入练习

1. 输入以下字母（左右手对应手指练习）

目标：录入速度达到 120 字符/分钟，即在 2 分钟以内完成录入。

vvmm	vvmm	mmvv	mmvv	vmvm	vmvm	mvmv	mvmv
bbnn	bbnn	nnbb	nnbb	bnbn	bnbn	nbnb	nbnb
vmbn	vmbn	vmnb	vmnb	mvbn	mvbn	mvnb	mvnb
bnvm	bnvm	bnmv	bnmv	nbvm	nbvm	nbmv	nbmv

vmfj　vmfj　mvjf　mvjf　bnfj　bnfj　nbjf　nbjf
vmgh　vmgh　mvhg　mvhg　bngh　bngh　nbhg　nbhg

2. 输入以下字母（强化练习）

目标：录入速度达到 90 字符/分钟，即在 5 分钟以内完成录入。

fvbm　jmnv　unbm　vtmb　nghv　mrvh
sonv　f;nv　vmws　nvoe　lnsv　jvim
sveb　ivob　mrsn　osnv　nmwv　vnsw
penv　nosm　ienv　;amv　nyub　tbnr
sivm　vjin　mvti　slmv　fvib　sebn
snvt　gvbf　hnmj　fmtn　ievb　nves
semb　ievm　vwen　sovn　nvje　mvni
sjnv　vein　mosb　ineb　imun　onrv
vrem　qbzn　ovpb　inwv　omvi　pnfv
ovin　envi　emvo　jnev　somv　pmln
dknv　almv　nkdv　lmcd　omrb　onsm
jnsv　lvbs　sdvn　lnmj　dnma　qbwv
wsvb　pnrm　okvn　knjb　;kvb　svkn
wkpb　lvmw　osvm　inem　sbkv　knsb
onkb　pm;v　onve　nivr　menw　avpm

3. 输入以下单词（综合练习）

目标：录入速度达到 90 字符/分钟，即在 3 分钟以内完成录入。

busy　busy　number　number　night　night
neither　neither　value　value　blue　blue
multiple　multiple　built　built　undue　undue
vacation　vacation　verb　verb　indent　indent
moment　moment　tempt　tempt　piano　piano
every　every　mother　mother　nine　nine
blurb　blurb　brave　brave　remain　remain
between　between　angry　angry　baby　baby
months　months　mountain　mountain　being　being

五、范围键【Z】【X】【C】【,】【.】【/】输入练习

1. 输入以下字母（左右手对应手指练习）

目标：录入速度达到 120 字符/分钟，即在 2 分钟以内完成录入。

zz//	zz//	//zz	//zz	z/z/	z/z/	/z/z	/z/z
xx..	xx..	..xx	..xx	x.x.	x.x.	.x.x	.x.x
cc,,	cc,,	,,cc	,,cc	c,c,	c,c,	,c,c	,c,c
z/a;	z/a;	a;z/	a;z/	/zpq	/zpq	pq/z	pq/z
x.wo	x.wo	ow.x	ow.x	x.sl	x.sl	ls.x	ls.x
c,dk	c,dk	kd,c	kd,c	c,dk	c,dk	kd,c	kd,c

2. 输入以下字母（强化练习）

目标：录入速度达到 90 字符/分钟，即在 5 分钟以内完成录入。

z/.x	x.z,	/.,z	xc.z	,c.x	.zcx
zcx/	.xz,	x/zc	cxz,	x.zc	.xc,
cd.l	lxmz	/slx	cl.s	xkrz	pzyx
pzwc	xeoz	oex.	j,sz	soxz	l.sz
evxc	idx,	/lyc	vhz,	colz	s,x/
slx,	a;cz	exrc	tcxz	p.ex	xvo,
zl/k	xs,l	expz	imx,	cei,	.zx,
id,z	li,m	rsvz	rks/	okdz	cvm,
sizc	epmx	xmzl	dciz	c,wo	.zxq
avcx	j;.,	ose,	zocq	rmxy	,ciz
x,cg	hzy/	iwc,	islx	irmc	;xow
cirw	urs.	xoe,	pdsz	w;xz	c,wp
xoem	,xpw	lasz	l,xe	iekz	coxn
vbxu	b,xe	zpq/	ktxi	comb	eiv.
cn.z	ei,c	xicm	b,zn	u,xe	wncv

3. 输入以下单词（综合练习）

目标：录入速度达到 90 字符/分钟，即在 3 分钟以内完成录入。

cut	cut	zest	zest	size	size
zero	zero	coin	coin	fox	fox
next	next	six	six	cold	cold
box	box	zoo	zoo	car	car
cox	cox	never	never	incline	incline
case	case	choose	choose	school	school

magic	magic	catch	catch	lazy	lazy
today.	today.	teacher.	teacher.	dish/	dish/
memory.	memory.	student.	student.	/photo	/photo

六、数字输入练习

1. 输入以下数字（左右手对应手指练习）

目标：录入速度达到 120 字符/分钟，即在 2 分钟以内完成录入。

1100	1100	0011	0011	1010	1010	0101	0101
2299	2299	9922	9922	2929	2929	9292	9292
3388	3388	8833	8833	3838	3838	8383	8383
4477	4477	7744	7744	4747	4747	7474	7474
5566	5566	6655	6655	5656	5656	6565	6565
4765	4765	9238	9238	1047	1047	6529	6529

2. 输入以下数字（强化练习）

目标：录入速度达到 90 字符/分钟，即在 5 分钟以内完成录入。

193845762	184759237	038572654
023858234	039438457	049375829
564376492	294829439	038428742
564783291	385712341	456823710
233925617	983746102	033826875
213945837	397546534	097806523
234985609	902478654	019849376
410435637	902381413	394856731
094557386	439285684	437579081
798076435	475256341	243425398
264353938	320530494	534849039
948528401	875617482	102938765
842104275	193758206	398576302
028760138	079834215	450922065
204839457	675321852	107492659

3. 输入以下语句（综合练习）

目标：录入速度达到 90 字符/分钟，即在 3 分钟以内完成录入。

1982 divided by 2 equals 991.

1982 divided by 2 equals 991.

3560 divided by 2 equals 1780.

3560 divided by 2 equals 1780.

1st. 2nd. 3rd. 4th. 72nd. 58th.

1st. 2nd. 3rd. 4th. 72nd. 58th.

18 times 93 equals 1674.

18 times 93 equals 1674.

256 times 381 equals 97536.

256 times 381 equals 97536.

灵丹妙药

一、遮盖键盘录入练习

盲打实际上是一种触觉打字，要用指尖的感觉去找键位，在练习过程中，要尽量避免不停地低头看键盘找键位。练习之初可以使用遮盖物覆盖在键盘上方，录入人员即使低头也看不见键盘，从而强迫其用手指去寻键。这样可能一开始速度会很慢，但只要坚持下去，就会很快渡过难关，并且中后期提速非常明显。

注意：遮盖物一定不能妨碍到录入人员双手的移动和击键，否则反而会影响正确指法的养成。

二、字母循环码录入练习

按照从 A～Z 的顺序输入 26 个字母，并反复多次循环输入，通过练习可以提高录入人员对于键位分布的掌握。

练习熟练后，再按照从 Z～A 的倒序输入 26 个字母，也需要反复多次循环输入。

三、手势还原练习

按照标准指法五要求，打字过程中要尽量遵循“手势还原”原则（每一根手指在移动击键的任务完成后，一定要习惯地回到基准键位置）。这是因为，从静止位置出发击打每一个键位，都只有唯一的一条移动路线，也就是指法唯一。通过练习，录入人员的手指很快就能记住这种指法，从而做到准确击键。而如果不做“手势还原”的动作，每次击键后直接移动到下一位置击键，则移动线路将变得复杂多变，很容易造成指法混乱，也就无法做到准确击键了。

可以练习在每次击打完一个范围键之后，立刻击打一次该范围键对应的基准键，从而强迫双手回到基准位。

注意：手势还原练习不要做得太多，以免养成范围键和基准键连击的习惯。

四、正确高效的姿势

要提高击键速度就要做到“悬腕”“立指”“贴浮”“节奏”。

● 悬腕：手腕不要压在或贴在键盘上，这样会影响录入人员双手移动的灵活性。

● 立指：手指的第一指节与键面尽量垂直，这样才能做到击键时以指尖垂直向字键使用冲力，瞬间发力并立即反弹，否则会变成按键或压键，速度和力量将下降很多。

● 贴浮：指尖贴近键面但不能触到，如果把手指放在字键上，很难发力；反之，如果手指离开键面的距离过远，明显增加了击键的键程，将会减慢速度。

● 节奏：连续击键不要一味求快，否则可能会降低准确率，最好的方式是能打出自己的节奏，在降低差错率的同时，不知不觉中速度就会提上来。

过关斩将

要求：反复练习以下每个部分的测试内容，必须在 5 分钟内输入完成（录入速度达到 120 字符/分钟），才能进入下一部分的测试，全部测试完成后，可以进入活动 2。

第一关：高考高频单词表录入

alter	burst	dispose	blast	consume	split	spit	spill
slip	slide	bacteria	breed	budget	candidate	campus	liberal
transform	transmit	transplant	transport	shift	vary	vanish	
swallow	suspicion	suspicious	mild	tender	nuisance		
insignificant	accelerate	absolute	boundary	brake	catalog		
vague	vain	extinct	extraordinary	extreme	agent	alcohol	
appeal	appreciate	approve	stimulate	acquire	accomplish		
network	tide	tidy	trace	torture	wander	wax	weave
preserve	abuse	academic	academy	battery	barrier	cargo	
career	vessel	vertical	oblige	obscure	extent	exterior	
external	petrol	petroleum	delay	decay	decent	route	

第二关：高频美语口语录入

I won't let her go without a fight.

It could happen to anyone.

I'm a laundry virgin.

I hear you.

Nothing to see here.

You are so sweet.

I think it works for me.

Rachel, you are out of my league.

You are so cute.

Let's get the exam rolling.

My way or the highway.

I planned to go there but something just came up.

That's not the point.

He shows up, we stick with him.

My life flashes before my eyes.

I have no idea what you have said.

Just follow my lead.

Good for you.

We're more than happy to give you recommendations.

We have to cut our trip short.

This party stinks.

You do the math.

I'm with you.

I will be there for you.

I would like to propose a toast.

第三关：高考作文模板录入

The topic of [ ] is becoming more and more popular recently. there are two sides of opinions of it. some people say a is their favorite. they hold their view for the reason of [ ]. what is more, [ ]. moreover, [ ].

While others think that b is a better choice in the following three reasons. firstly, [ ]. secondly, [ ]. thirdly, [ ].

From my point of view, I think [ ]. the reason is that [ ]. as a matter of fact, there are some other reasons to explain my choice. for me, the former is surely a wise choice.

Some people believe that [ ]. for example, they think [ ]. and it will bring them [ ].

In my opinion,I never think this reason can be the point. for one thing,[ ]. for another thing,[ ].

Form all what I have said,I agree to the thought that [ ].

活动 2　百发百中——纠正指法

经过一段时间的刻苦练习，小文和小璐终于连闯三关，通过了活动 1 的所有测试环节。尤其是小璐，她改变了自己以前习惯的“一指禅”，现在可以十指翻飞地打字了。他俩心理颇有几分得意，都急着想学点新本领了。

可师傅却不紧不慢地说：“让我先来看看你们这段时间的练习成果吧。”

随着一声令下，小文和小璐开始录入一篇近千字符的英文文章。不到 6 分钟，他俩就都完成了。可当他们满怀期待地等着师傅的夸奖时，师傅却笑呵呵地冲他们伸出了右手小拇指，还打趣地说：“都生老茧了吧！”

就在小文和小璐一头雾水的时候，师傅严肃地说道：“你们在追求速度的同时，却忽略了文录的另一个重要的指标——准确率！在你们刚才打字的过程中，由于击键的准确率不够，所以频繁使用回退键【Backspace】回退修改。要知道打错一个字符，就必须多打一次回退键【Backspace】，再重新输入正确的字符。也就是说，原本击键一次的事现在至少要击键三次才能完成！”

“不断的修改还会打断录入人员的思路，对输入速度带来致命的影响。”看看两个有点不知所措的徒弟，师傅继续说道，“别急，下一步我就来教你们怎样降低回改率（回退键【Backspace】的使用频率），提高准确率。这样，你们的录入速度也会有一个新的飞跃。”

摩拳擦掌

首先，我们必须找准问题，搞清楚自己有哪些键容易打错。下面介绍一种简单易行的“黑屏”查错法。

第一步，在电脑上打开任意一种文本编辑软件，比如 Word、记事本或写字板，新建一个空白文档。

第二步，非常重要！关闭电脑显示器！也就是说，下面你将面对一块“黑屏”录入字符。

第三步，盲打录入以下固定短句，每句话都要连续录入十遍以上，每录入完成一遍就敲一次回车键【Enter】换一行。

- Abcdefg hijklmn opq rst uvw xyz
- Where there is a will there is a way
- All work and no play makes jack a dull boy
- Every man has his faults
- Failure is the mother of success
- Where there is life there is hope
- You cannot eat your cake and have it
- We shall never have friends if we expect to find them without fault
- Unity is strength
- A man cannot spin and reel at the same time
- A friend in need is a friend indeed
- A young idler an old beggar
- A little knowledge is a dangerous thing
- An early bird catches the worms
- An hour in the morning is worth two in the evening
- Every advantage has its disadvantage
- A good beginning is half done
- Do nothing by halves

第四步，打开显示器，仔细查对录入中的错误，特别是反复出现的错误，并保存记录下来。这些就是你指法中存在的问题了。

以上方法，在练习文录的任何阶段，都可以作为自我检查的有效办法。

师傅领进门

对于初学者常见的指法问题，主要分为以下几种：

一、基准键错位

如果输入的结果出现连续的大面积错误，一般是由于一次移动击键的任务完成后，手指没有回到正确的基准键位置上，而从错误的基准键位置出发再去击键，后面的字符就全都错位了，如图 1—2—1 所示。

图 1—2—1　基准键错位

此类错误要先养成依靠触摸盲打坐标（【F】和【J】键上的小突起）定位基准键的习惯，然后加强手势还原练习，一般很快就可以克服。

二、左右不分

就是把一只手负责的范围键，打成了另一只手相应手指的范围键，比如【B】和【N】打错，【Q】和【P】打错，【S】和【L】打错等，如图 1—2—2 所示。

此类错误，一部分是因为键位排列不够熟练，可以做一些单手录入练习，强化左右手分区键位记忆；另一部分则是因为初学阶段过于追求速度，导致瞬间反应跟不上，这种情况依靠单手录入练习是不解决问题的，必须先把录入速度降下来，争取在保证准确率的基础上打出节奏，再逐渐提速。

图 1—2—2　左右手范围键颠倒

三、上下不分

就是一根手指负责的上排键、中排键、下排键打乱，比如【E】【D】和【C】打错，【U】【J】和【M】打错等，如图 1—2—3 所示。

此类错误主要是因为对键位排列不够熟悉，可以通过做一些单排键录入练习来加以改善。但需要注意的是，做单排键录入练习时也要手势还原，切不可就把手停在练习录入的那排键上。

图 1—2—3　上排键、中排键、下排键打乱

四、字符重复

如果发现某些字符被错误地连续录入，一般是因为击键的手势不对，手指摆放得太平或击键的力量太大，就容易把击键变成了按键或压键，所以一次按键导致重复输入多个字符。想避免此类错误就要尽量做到“立指”，就是手指的第一指节与键面尽量垂直，这样就能在击键时以指尖垂直向字键使用冲力，瞬间发力并立即反弹。

五、字符遗漏

就是应该录入的字符没有录进去。此类错误一般多见于小指管辖区域，大多是因为小指力量不足，在高速打字时虽然做出了击键动作，但字键并没有被按下，所以字符没有录入。只要平时加强小指的力量练习，比如空闲时在桌面或大腿上做击打练习，就能获得改善。

六、字符添加

就是不应该录入的字符多输入了。此类错误一般多见于无名指管辖区域，由于人们对无名指的控制力偏弱，用无名指击键时可能会连带到中指或小指，所以相邻的两个字符会一起被录入。只要平时加强无名指的控制练习，比如空闲时在桌面或大腿上做击打练习，就能获得改善。

还有一类是多录入空格，这主要是因为录入人员把大拇指放在空格键上，在录入其他字符时就容易带到空格。此类错误只要注意适当调整大拇指位置，做到“贴浮”（手指贴近键面但不接触）就可以避免。

七、字符颠倒

就是把字符的排列顺序录入颠倒，比如把“dir”打成“dri”或“idr”。此类错误多是因为盲目求快造成的，还是要打出适合自己的“节奏”来，不知不觉中错误少了，录入速度就会提高。

八、个别键定位不准

如果发现自己在录入中有个别字符经常出错，那就需要特别注意了，这往往是因为个人的习惯性指法不够标准，范围键分工不明造成的。此类错误一定要通过针对性强化练习加以克服，一个一个地把不规范指法矫正过来，并形成习惯。

对于自身存在的指法问题，一定要对症下药，尽早改正，以免长期使用下来，错误的习惯反而被固化下来，以后就很难再改了。特别需要注意的是，在纠正指法的过程中一定要坚持盲打，千万不能依靠视觉来纠错！

修行靠个人

以下提供一些针对性训练，可以根据自我检查的结果来选择练习。其中，最后两项练习是必做的！全部练习完成后，可以再次使用“黑屏”查错法检查训练效果，对不足之处要反复练习。

一、手势还原练习

目标：录入速度达到150字符/分钟，即在3分钟以内完成录入。

altfedrf bfujrfstf diksp；olsed bflastf cdolnjsujmjed
sp；liktf slikded bfacdtfedrfika bfrfededd bfujdgfedtf
cdanjdikdatfed cdamjp；ujs likbfedrfal tfrfanjsfolrfmj
shjikftf vfarfyj vfanjikshj swsallolws tfednjdedrf
sujsp；ikcdikolnj mjikld njujiksanjcded acdcdedledrfatfed
abfsollujtfed bfolujnjdarfyj bfrfaked cdatfalolgf
vfagfujed vfaiknj edxstfiknjcdtf edxstfrfedmjed
agfednjtf alcdolhjoll ap；p；edal acdqaujikrfed
ap；p；rfolvfed stfikmjujlatfed ap；p；rfedcdikatfed

二、单手录入练习

1. 左手录入练习

目标：录入速度达到150字符/分钟，即在3分钟以内完成录入。

tgfdrwaz bvfdcsr tgfrbde qwsxdaf zewxsda crsexada
gtfdread vxtgdrf desxcaw wdeaqvb gtxzdrd cxqtcaze
dsgrevxe sdwxaqt gbrsbaq evgsdxt awveqzt gztqrwdx
tzwbrqfa vcwqzrx csedtab wvsrqze wsaddxa zrtwqsaq
drtgcvxs aetargd bxsetqz fsdsgxc azsewgb fdvarxdw
gxsztqvb tagzbqr svwfxeg acdtqbz drsrtag wrxvefsd
tqgadbxz rwvdfse tdegbxd earqfdf gxcdsse bdrxasqz
gtfrgbvf dtgsetg aeewvbz zddcset rgdsxqt sdeavtgs
xtdrefgw vxtgedr seavvxb zqwsxtg swbedra vqfzedsa
erxcdeet gbdrxst grfbsrx defgcxz rtgfaqg swzgbxsq

2. 右手录入练习

目标：录入速度达到150字符/分钟，即在3分钟以内完成录入。

yjikl/mn polikhj yipkjnl uhkl,jh kh. njik ol;kijym
kiujhm,/ jikulhj ol,mnoh polikny ynhujmi olkijhu;
,ijuhulo opol. in mko,hyi unhikji klji;oh y;po. kim
imjholku nlo;jku ol;pjiu nhio;lk okilok; ikolk;jy

jikmnuop	l,jinm;	hji,uno	ploikni	uohymn;	. loijlun
uykilkj/	oiukmnh	olkjinl	olmnkhy	hikjlom	ikjuhnm.
huijkmin	ykol,mn	jiop;lk	hilkmjy	nop;ymk	l;jkmopy
loikjmnu	luinhyk	mkijolp	nholpmy	onikyjh	iylomnju
plijhy,n	yipl,mj	kinjuh;	nolk,pj	hikjuol	plk,mnik
yjio;lkn	olkim;h	kio,mhj	opluynh	jiolkph	njimkolp

三、单排键录入练习

1. 上排键录入练习

目标：录入速度达到150字符/分钟，即在3分钟以内完成录入。

will hold pass look park pull swell told world follow the path as far as it goes quite short that the jewels are our property equal quote fought just request weight post poster post good deal a lot of a pair of add to allow for as follows for if we to as well day after die away dress up equal to fall away fall ill first free of get up go away go through keep off last of all lead to look after look for like out of date play house so far as a good add keep late out ok date play do far

2. 下排键录入练习

目标：录入速度达到150字符/分钟，即在3分钟以内完成录入。

small bkxd gnazc mfnc caxxjz ndf xjk danc vanms zjlk kncj cjfn lsk cfc nfj xkd xl-la mncjd nj cdvf sxjkz almn df jxn csfjg cnj dkcz slam xnjsd jd cnjd slxk nxm nfj kdnc jf xjs zjfdal dkcjf mkcf kmnx jfd dms sjfd nmxk dkmxd jdmx nbjh jnv nvgfk smjxfn bjxl djmx nvf axz lmcjf ckdng xjdnms bzjdn skx zmdn nn xjdmzd bvchv dkxm nvs jskd mxnj lx ndfmkjx dmslkx kcmfj ndlxm jnhmk kmnjxh fbscnb bhs xxjk ndhcb jakmxxn bdj mzkvn shvxm nnf dgb vh cvnxm mxvj fjszs bvcx kdmvx mxzjf dnv msnx kchn dbk

四、个别键针对性强化练习

本练习内容要根据个人的情况进行设定。比如：某人在录入过程中经常把【R】错打成【T】的，就可以进行以下内容的录入练习。

目标：录入速度达到120字符/分钟，即在5分钟以内完成录入。

rabbit	race	radio	railroad	rain	rainbow	rainy	raise
rat	rather	reach	read	ready	real	reason	receive
red	refrigerator	remember	repair	repeat	restaurant	rest	return
ribbon	rice	rich	ride	right	river	dry	ring

road	room	rose	round	row	dark	after	agree
run	air	angry	april	arrive	art	bear	bird
borrow	dream	break	bridge	brown	brush	bring	burn
butter	carry	cry	card	careful	carrot	chair	circle
clear	color	corner	drive	drink	earth	february	first
four	free	frog	friend	girl	garden	grade	great
green	group	grow	hair	heart	here	horse	hurry
hungry	iron	large	later	learn	letter	library	march
mark	marry	mirror	merry	more	near	nor	narrow
never	orange	order	organ	over	pair	paper	pardon
parent	our	part	pearl	person	poor	power	prepare

五、标准指法巩固练习

1. 练习一：单词录入

目标：录入速度达到120字符/分钟，即在5分钟以内完成录入。

satellite	ruin	sake	scale	temple	tedious	tend
tendency	ultimate	undergo	abundant	adopt	adapt	bachelor
casual	trap	vacant	vacuum	oral	expenditure	optics
organ	excess	expel	expend	individual	expense	expensive
expand	expansion	private	personal	personnel	ocean	coil
continual	acknowledge	balcony	grant	grand	invade	acid
calculate	calendar	optimistic	optional	outstanding	religious	export
import	impose	religion	victim	videotape	video	offend
bother	radical	hook	beforehand	internal	racial	radiation
ban	coach	adequate	range	wonder	isolate	issue
hollow	consistent	coarse	valid	valley	adhere	capture

2. 练习二：短句录入

目标：录入速度达到120字符/分钟，即在5分钟以内完成录入。

Live beautifully, dream passionately, love completely.

A man's best friends are his ten fingers.

All things in their being are good for something.

Failure is the mother of success.

The unexamined life is not worth living.

Suffering is the most powerful teacher of life.

Living without an aim is like sailing without a compass.

Bad times make a good man.

A man can't ride your back unless it is bent.

When all else is lost the future still remains.

The world is his who enjoys it.

Towering genius disdains a beaten path.

It seeks regions hitherto unexplored.

Victory belongs to the most persevering.

Genius is an infinite capacity for taking pains.

Adversity is the midwife of genius.

3. 练习三：文章录入

目标：录入速度达到 120 字符/分钟，即在 5 分钟以内完成录入。

My father was a self-taught mandolin player. He was one of the best string instrument players in our town. He could not read music, but if he heard a tune a few times, he could play it. When he was younger, he was a member of a small country music band. They would play at local dances and on a few occasions would play for the local radio station. He often told us how he had auditioned and earned a position in a band that featured Patsy Cline as their lead singer. He told the family that after he was hired he never went back. Dad was a very religious man. He stated that there was a lot of drinking and cursing the day of his audition and he did not want to be around that type of environment.

灵丹妙药

一、使用“打字旋风”软件练习录入

打字练习软件“打字旋风”带有回退键【Backspace】使用次数统计的功能。在软件中练习输入，完成后进入查看成绩界面，就能够看到如图 1—2—4 所示的统计表。

练习模式	速度	正确率	正确	错误	回退	总键数	总字符数	练习时间(秒)
英文文章...	43.48字/分	100.00%	20	0	9	39	20	27

图 1—2—4　打字数据统计表

其中“正确”“回退”“总键数”和“总字符数”反映了录入人员的指法准确程度（回退率）和击键有效率。比如图 1—2—4 中显示的录入结果：此次共录入了 20 个字符，虽然最终 20 个都正确，但录入过程中击键总数却达到 39 下，击键有效率＝正确字符数/总键数＝20/39＝51％，回退率＝回退数/总字符数＝9/20＝45％。

很明显，过多地使用回退键【Backspace】，会极大地降低击键有效率，速度也就大打折扣了。而提高指法准确程度，降低回改率，就可以提高击键有效率，录入速度也就相应提高了。

在练习录入的过程中使用“打字旋风”软件，经常关注以上数据，并不断使用上面介绍的五种练习方法纠正指法，争取把回改率控制在 10％以下，击键有效率保持在 80％以上，录入速度就会大幅增长。

二、使用“打字高手”软件练习录入

打字练习软件“打字高手”带有大量的指法训练模块。虽然对于“范围键”的分区练习和本书活动 1 介绍的分区方法不完全一致，但作为指法巩固练习却是非常方便的。该软件还提供了“漫游迷宫”“圣诞礼物”和“采蘑菇”三个打字游戏，可以增进练习打字的兴趣，如图 1—2—5所示。

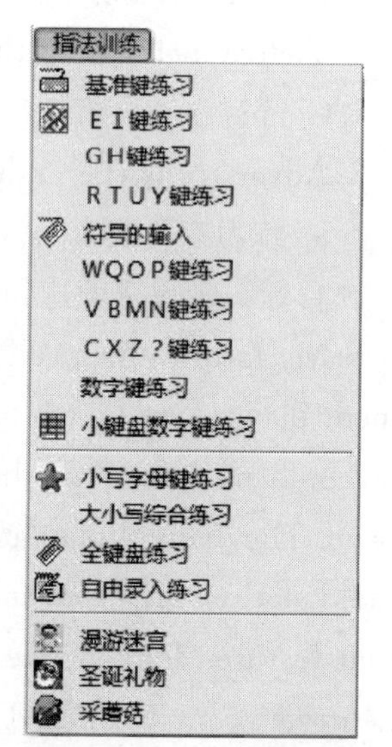

图 1—2—5 打字游戏

过关斩将

要求：反复练习以下每个部分的测试内容，必须在 5 分钟内输入完成（录入速度达到 150 字符/分钟），才能进入下一部分的测试，全部测试完成后，可以进入活动 3。

第一关：高考高频单词表录入

continuous	interfere	explode	exploit	explore	remote
explosion	render	removal	explosive	identify	idle
precaution	poverty	resolve	barrel	bargain	code
advertisement	resistant	advertise	adult	agency	focus
forbid	debate	debt	decade	enclose	encounter
globe	global	scan	scandal	significance	subsequent
virtue	virtual	portion	target	portable	decline

illusion	likelihood	stripe	emphasize	emotion	emotional
awful	awkward	clue	collision	device	devise
inevitable	naval	navigation	necessity	previous	provision
pursue	stale	substitute	deserve	discrimination	professional
secure	security	scratch	talent	insurance	insure
nevertheless	neutral	spot	spray	medium	media
auxiliary	automatic	compete	distribute	disturb	infer
cancel	integrate	moist	moisture	promote	region
register	stable	sophisticated	splendid	variable	prospect

第二关：名言、俗语录入

The word impossible is not in my dictionary.

An aim in life is the only fortune worth finding.

While there is life there is hope.

He who fears being conquered is sure of defeat.

Take time to deliberate, but when the time for action has arrived, stop thinking and go.

Nothing is more difficult, and therefore more precious, than to be able to decide.

A people which is able to say everything becomes able to do everything.

You have to believe in yourself. that is the secret of success.

Pursue your object, be it what it will, steadily and indefatigably.

Courage is like love; it must have hope to nourish it.

We must accept finite disappointment, but we must never lose infinite hope.

Other men live to eat, while i eat to live.

Energy and persistence conquer all things.

Cease to struggle and you cease to live.

A thousand-li journey is started by taking the first step.

第三关：美文录入

Love is something we all need. but how do we know when we experience it?

True love is best seen as the promotion and action, not an emotion. love is not exclusively based how we feel. certainly our emotions are involved. but they cannot be our only criteria for love. true love is when you care enough about another person that you will lay down your life for them. when this happens, then love truly is as strong as death. how many of you have a mother, or father, husband or wife, son or daughter or friend who would sacrifice his or her own life on yours? those of you who truly love your spells but unchildren, would unselfishly lay your life on the line to save them from death? many people in an emergency room with their loved ones and prayed "please, god, take me instead of them". find true love and be a true lover as well. may you find a love which is not only strong as death, but to leave to a truly for feeling life.

活动3　运指如飞——文章练习

经过上一阶段的自我检查和针对性训练，小文和小璐的指法得到了纠正和巩固。虽然反复的练习有些枯燥，但令他们惊喜的是：随着回退键【Backspace】使用量的减少，两人的录入速度都有非常明显的提升，特别是打字录入时手上不紧不慢，而字符却一行行飞快地跃上屏幕的感觉真好！

师傅告诉他们：这就是文录的节奏。一个打字高手，你看他录入时并不会觉得他的手指运动有多快，因为他的每一下动作都是准确、简捷、高效的，没有多余花哨的动作。

最重要的是：错误百出的高速打字和百分百准确的低速打字都是没有实际意义的！我们学习文录，就是要做到准确与速度并重，并且从练习时就要二者兼顾。其实，这二者也是相辅相成的，在活动 2 中，准确性对于速度提升的意义相信大家都有体会。在活动 3 中，我们将继续提速，看看谁先成为当之无愧的打字高手！

摩拳擦掌

在录入英文的实际应用中，无论是单词中的专用名词，还是句子中的首字母，都必然会遇到大小写字母的混合录入问题。下面，首先来认识一下用于英文大小写切换的功能键：

● 大小写字母锁定键【Caps Lock】：大小写锁定键是一个开关键。按下该键，键盘上的 Caps Lock 指示灯会亮，表示当前处在大写锁定状态，此时输入的所有英文字符均为大写；再按一下该键，Caps Lock 指示灯熄灭，表示当前处在小写锁定状态，输入的所有英文字符均为小写。

● 换挡键【Shift】：换挡键的一个功能就是进行大小写的转换。在小写锁定状态，按住换挡键【Shift】的同时再按字母键，输入相应的大写字母；反之，在大写锁定状态，按住换挡键【Shift】的同时再按字母键，输入相应的小写字母。

换挡键的另一个功能是输入双挡键的上挡字符（键盘上有些按键的键面上标有一上一下两个字符，这样的键叫双挡键，其中上面的叫上挡字符，下面的叫下挡字符）。在不按换挡键【Shift】时，击打双挡键输入下挡字符；按下换挡键【Shift】时，击打双挡键输入上档字符。

师傅领进门

通过使用换挡键【Shift】和大小写字母锁定键【Caps Lock】，就能够达到大小写字母混合录入的要求，标准指法如下：

一、需用左手录入的大写字母，要用右手小指按右下角的换挡键【Shift】；需用右手录入的大写字母，要用左手小指按左下角的换挡键【Shift】。这就是说，大写英文字母录入一定要左右手配合完成，才是高速正确的指法！

二、要先按下换挡键【Shift】，再击打需要输入的字母键，然后双手可以同时离开键面。

三、如果需要连续录入大写字母时，先用左手小指击打大小写字母锁定键【Caps Lock】，然后就像录入小写字母一样击键打字；等大写字母录入完成后，再次用左手小指击打大小写字母锁定键【Caps Lock】，恢复小写字母录入状态。

注：双挡键的上挡字符录入的标准指法同上一、二。

修行靠个人

一、大小写字母混合录入

1. 专用名词录入练习

目标：录入速度达到 120 字符/分钟，即在 5 分钟以内完成录入。

Jan. January	Feb. February	Mar. March	Apr. April
May	Jun. June	Jul. July	Aug. August
Sep. September	Oct. October	Nov. November	Dec. December
Mon. Monday	Tue. Tuesday	Wed. Wednesday	Thu. Thursday
Fri. Friday	Sat. Saturday	Sun. Sunday	

The Forbidden City	The Ming Tombs	The Nine Dragon Screen	
The Temple of Heaven	The Summer Palace	Huaqing Hot Spring	
The 17-Arch Bridge	The Marble Boat	The Great Wall	
the Palace Museum	the Meridian Gate	Longmen Cave	
Lushan Mountain	Heaven Poll	Big Wild Goose Pagoda	
Huashan Mountain	Emei Mountain	Stone Forest	Yueyang Tower
West Lake	White Horse Temple	Potala Palace	Echo Wall
Grand Canal	Du Fu Cottage	Dujiang Dam	Gulangyu Islet

2. 短句录入练习

目标：录入速度达到 120 字符/分钟，即在 5 分钟以内完成录入。

It's up to you.

I envy you.

How can I get in touch with you?

Where can I wash my hands?

What's the weather like today?

Where are you heading?

I wasn't born yesterday.

What do you do for relaxation?

It's a small world.

It's my treat this time.

The sooner the better.

When is the most convenient time for you?

Take your time.

I'm mad about Bruce Lee.

I'm crazy about rock music.

How do I address you?

What was your name again?

Would you care for a cup of coffee?

She turns me off.

So far so good.

It drives me crazy.

She never showed up.

That's not like him.

I couldn't get through.

I got sick and tired of hotels.

Be my guest.

Can you keep an eye on my bag?

Let's keep in touch.

Let's call it a day.

I couldn't help it.

Something's coming up.

3. 文章录入练习

目标：录入速度达到120字符/分钟，即在5分钟以内完成录入。

The Fox And The Cock

One morning a fox saw a cock. He thought,"This is my breakfast. "He came up to the cock and said,"I know you can sing very well. Can you sing for me?" The cock was glad. He closes his eyes and began to sing. The fox saw that and caught him in his mouth and carried him away.

The people in the field saw the fox. They cried,"Look,look! The fox is carrying the cock away. " The cock said to the fox,"Mr. Fox,do you understand? The people say you are carrying their cock away. Tell them it is yours. Not theirs. "

The fox opened his mouth and said,"The cock is mine,not yours. " Just then the cock ran away from the fox and fled into the tree.

二、英文录入提速

1. 专用名词录入练习

目标：录入速度达到150字符/分钟，即在5分钟以内完成录入。

Fortune　Desk Research　Ogilvy & Mather　White Goods　Infocus

Budweiser　Banded Pack　Procter & Gamble　Product Line　Polaroid

Marginal Cost　Marginal Benefit　Standard Deviation　Buick　BMW

Product Mix　Presentation　Supermarket　Orange Goods　Sampling

Pyramid Selling　Pass-on Circulation　Word Associaton　Promotion

DavidOgilvy　DeBeers　Elasticity　Reach　Dodge　Topof Mind

Telephone Interview　FIAT Readers　FUJIFILM　Dentsu　DuPont

Secondary Data　Circulation　MeToo　Interview　Interviewer　ABC

Philip Morris Company　EPOS Data　IKEA　Soap Opera　Split-runTest
Distribution　TOYATA　Gap Analysis　Floating Spot　Aided Recall
Paid Circulation　Back Checking　Validation　VOLVO　Parliament
Kodak　Shell　Variable Pricing　Coca-Cola　Kleenex　Believability
Chrysler　Claude Hopkins　Outside View　Kentucky Fried Chicken
OTC　SIEMENS　VISA　MasterCard　General Motor　Time Warner

2. 短句录入练习

目标：录入速度达到150字符/分钟，即在5分钟以内完成录入。

Let's get to the point.
Keep that in mind.
That was a close call.
I'll be looking forward to it.
Chances are slim.
Far from it.
I'm behind in my work.
It's a pain in the neck.
We're in the same boat.
My mouth is watering.
What do you recommend?
I ache all over.
I have a runny nose.
It's out of the question.
Do you have any openings?
It doesn't make any difference.
I'm fed up with him.
You can count on us.
It doesn't work.
It's better than nothing.
Think nothing of it.
I'm not myself today.
I have a sweet tooth.
I can't express myself very well in English.

For the time being.

This milk has gone bad.

Don't beat around the bush.

It's up in the air.

Math is beyond me.

It slipped my mind.

You can't please everyone.

I'm working on it.

You bet!

Drop me a line.

Are you pulling my leg?

Sooner or later.

I'll keep my ears open.

It isn't much.

Neck and neck.

I'm feeling under the weather.

3. 文章录入练习

目标：录入速度达到 150 字符/分钟，即在 5 分钟以内完成录入。

In August of 1993 my father was diagnosed with inoperable lung cancer. He chose not to receive chemotherapy treatments so that he could live out the rest of his life in dignity. About a week before his death, we asked Dad if he would play the mandolin for us. He made excuses but said "okay". He knew it would probably be the last time he would play for us. He tuned up the old mandolin and played a few notes. When I looked around, there was not a dry eye in the family. We saw before us a quiet humble man with an inner strength that comes from knowing God, and living with him in one's life. Dad would never play the mandolin for us again. We felt at the time that he wouldn't have enough strength to play, and that makes the memory of that day even stronger. Dad was doing something he had done all his life, giving. As sick as he was, he was still pleasing others. Dad sure could play that Mandolin!

三、综合应用

目标：录入速度达到 150 字符/分钟，即在 10 分钟以内完成录入。

文章内容	累计字数
Occasionally, Dad would get out his mandolin and play for	58
the family. We three children: Trisha, Monte and I, George Jr. ,	122
would often sing along. Songs such as the Tennessee Waltz,	181
Harbor Lights and around Christmas time, the well-known	237
rendition of Silver Bells. " Silver Bells, Silver Bells, its	297
Christmas time in the city" would ring throughout the house.	358
One of Dad's favorite hymns was "The Old Rugged Cross". We	417
learned the words to the hymn when we were very young, and would	482
sing it with Dad when he would play and sing. Another song that	546
was often shared in our house was a song that accompanied the	608
Walt Disney series: Davey Crockett. Dad only had to hear the	669
song twice before he learned it well enough to play it. "Davey,	733
Davey Crockett, King of the Wild Frontier" was a favorite song	796
for the family. He knew we enjoyed the song and the program	856
and would often get out the mandolin after the program was	915
over. I could never get over how he could play the songs so	975
well after only hearing them a few times. I loved to sing, but	1038
I never learned how to play the mandolin. This is something	1098
I regret to this day.	1119
Dad loved to play the mandolin for his family he knew we	1176
enjoyed singing, and hearing him play. He was like that. If	1236
he could give pleasure to others, he would, especially his	1295
family. He was always there, sacrificing his time and efforts	1357
to see that his family had enough in their life. I had to mature	1422
into a man and have children of my own before I realized how	1483
much he had sacrificed.	1506

按照10分钟的时限进行以上练习，每次完成后记录下累计录入的总字数，通过反复练习直至达到既定目标。

以下内容用于巩固练习，可为自己设定一个较高的目标，并通过反复练习达成目标。200字符/分钟的录入速度是每个人都能够做到的。

文章内容	累计字数
I joined the United States Air Force in January of 1962.	57
Whenever I would come home on leave, I would ask Dad to play	118
the mandolin. Nobody played the mandolin like my father. He	178
could touch your soul with the tones that came out of that old	241
mandolin. He seemed to shine when he was playing. You could	301
see his pride in his ability to play so well for his family.	361
When Dad was younger, he worked for his father on the farm.	421
His father was a farmer and sharecropped a farm for the man	481
who owned the property. In 1950, our family moved from the	540
farm. Dad had gained employment at the local limestone quarry.	603
When the quarry closed in August of 1957, he had to seek other	666
employment. He worked for Owens Yacht Company in Dundalk,	724
Maryland and for Todd Steel in Point of Rocks, Maryland. While	787
working at Todd Steel, he was involved in an accident. His job	850
was to roll angle iron onto a conveyor so that the welders	909
farther up the production line would have it to complete their	972
job. On this particular day Dad got the third index finger of	1034
his left hand mashed between two pieces of steel. The doctor	1095
who operated on the finger could not save it, and Dad ended	1155
up having the tip of the finger amputated. He didn't lose	1213
enough of the finger where it would stop him picking up	1269
anything, but it did impact his ability to play the mandolin.	1330
After the accident, Dad was reluctant to play the	1380
mandolin. He felt that he could not play as well as he had	1439
before the accident. When I came home on leave and asked him	1500
to play he would make excuses for why he couldn't play.	1556
Eventually, we would wear him down and he would say "Okay, but	1619
remember, I can't hold down on the strings the way I used to"	1681
or "Since the accident to this finger I can't play as good".	1742
For the family it didn't make any difference that Dad couldn't	1805
play as well. We were just glad that he would play. When he	1865
played the old mandolin it would carry us back to a cheerful,	1927
happier time in our lives. "Davey, Davey Crockett, King of the	1990
Wild Frontier", would again be heard in the little town of	2049
Bakerton, West Virginia.	2073

灵丹妙药

要进一步提高录入人员的英文打字速度，除了通过练习不断提升手指击键的速率，以及击键的准确度以外，还要注意以下一些细节，只有养成良好的习惯，才能成为真正的打字高手。

一、交替使用双手敲击空格键

由于在实际应用中，空格键是使用频率最高的按键（占英文文章字符总数的近20%），在输入空格时轮换使用左、右手的大拇指来敲击，就不会造成单根手指过度疲劳，达到提速的目的。

二、大拇指尽量靠近空格键的中部击键

由于空格键很长，这虽然给击键带来了方便，但如果长期击打两端的位置，很容易造成按键的接触不良，就会造成击键无效、延迟或连续空格的错误，对录入速度有很大影响。正确的手势是双手大拇指自然舒展开，靠近空格键的中部摆放，而不要蜷缩在食指的下方。

三、双手配合使用

打字时有时需要同时按下两个键，如果这两个键分布于左右两个区，则一定用左、右手各击其键。特别需要注意的是：【Shift】【Ctrl】【Alt】键都是左右两区各有一个，空格键虽然只有一个，但同时跨越左右两区，在使用时要合理选择一区使用，以保证双手的配合。

四、良好的坐姿和手势

打字过程中错误的坐姿和手势容易造成录入人员的疲劳，影响文录的高速稳定。一个真正优秀的录入员，在开始工作前，会很仔细地根据桌面调整座椅的高度、键盘的位置、屏幕与眼睛的距离、录入文稿的摆放等，如图 1—3—1 所示的标准坐姿仅供参考。

五、击键力度适中

打字时敲击键盘的力量要适度，击键太轻容易造成漏码，而击键太重不仅容易造成重码，还会毁坏键盘。对于不同的键盘，手感也不完全一样。一般的键盘和笔记本电脑键盘，以及专用的速录键盘，它们的按键冲程不同，击键的最佳力度也不一样。

六、反复练习同一段文稿

当你练习打字陷入困境，速度提升遭遇瓶颈时，有效的解决办法之一就是反复练习同一段文稿。每次练习要记录速度，并总结一下经常出错的环节，然后再练，直至

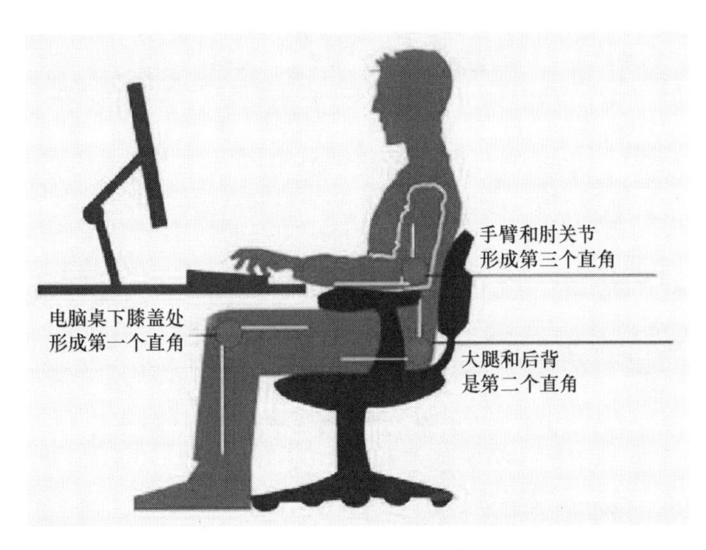

图 1—3—1　标准坐姿

达到既定的速度和准确性再换稿。千万不要贪多，文稿打一遍就换，看似练了很多，但缺乏针对性，练习效率低下。

七、特殊指法："手势连动还原"原则

在标准指法中，打字过程要遵循"手势还原"原则（每一根手指在移动击键的任务完成后，一定回到基准键位置）。但随着录入员对键盘熟悉程度的提高，可以尝试升级到"手势连动还原"，即遇到两个及以上的连续字符是用同一根手指击键输入时，可以连续移动该手指击键，全部完成后再回到基准键位置，这样速度更快。

例如：输入 number，【n】【u】【m】三个连续字符键都是右手食指的范围键，两种指法的比较如图 1—3—2 所示。

八、特殊指法：手指的"平行"移动原则

在标准指法中，打字时手指要遵循"独立"移动原则（一根手指在击键过程中独立运动，其余不击键的手指不要跟着一起离开基准键位）。在练习录入的高级阶段，可以尝试升级到"平行"移动原则，即遇到两个及以上的连续字符，是可以用单手的多根手指在同一排键位上输入的，则全手多根手指平行移动击键，全部完成后再一起回到基准键位置，这样速度更快。

例如：输入 were，【w】【e】【r】【e】四个连续字符键都是左手的上排键，两种指法的比较如图 1—3—3 所示。

图 1—3—2 “手势还原”与“手势连动还原”指法比较

图 1—3—3 “独立移动”与“平行移动”指法比较

过关斩将

要求：反复练习以下每个部分的测试内容，必须在 5 分钟内输入完成（录入速度达到 180 字符/分钟），才能进入下一部分的测试，全部测试完成后，可以进入后续活动。

第一关：英文小故事录入

Once there were two mice. They were friends. One mouse lived in the country;the other mouse lived in the city. After many years the Country mouse saw the City mouse; he said,"Do come and see me at my house in the country." So the City mouse went. The City mouse said,"This food is not good,and your house is not good. Why do you live in a hole in the field? You should come and live in the city. You would live in a nice house made of stone. You would have nice food to eat. You must come and see me at my house in the city."

The Country mouse went to the house of the City mouse. It was a very good house. Nice food was set ready for them to eat. But just as they began to eat they heard a great noise. The City mouse cried," Run! Run! The cat is coming!" They ran away quickly and hid.

After some time they came out. When they came out,the Country mouse said,"I do not like living in the city. I like living in my hole in the field. For it is nicer to be poor and happy,than to be rich and afraid."

第二关：英文演讲稿录入

Youth is not a time of life; it is a state of mind; it is not a matter of rosy cheeks, red lips and supple knees; it is a matter of the will,a quality of the imagination,a vigor of the emotions; it is the freshness of the deep springs of life.

Youth means a temperamental predominance of courage over timidity of the appetite for adventure over the love of ease. This often exists in a man of 60 more than a boy of 20. Nobody grows old merely by a number of years. We grow old by deserting our ideals.

Years may wrinkle the skin,but to give up enthusiasm wrinkles the soul. Worry, fear,self-distrust bows the heart and turns the spirit back to dust.

Whether 60 of 16,there is in every human being's heart,the lure of wonders,the unfailing childlike appetite of what's next and the joy of the game of living. In the center of your heart and my heart there's a wireless station: So long as it receives messages of beauty,hope,cheer,courage and power from men and from the infinite,so long as you are young.

第三关：英文经典美文录入（上篇）

Once upon a time there was a child ready to be born. So one day he asked God, "They tell me you are sending me to earth tomorrow but how am I going to live there being so small and helpless?"

God replied,"Among the many angels,I chose one for you. She will be waiting for you and will take care of you."

But the child wasn't sure he really wanted to go. "But tell me, here in Heaven, I don't do anything else but sing and smile, that's enough for me to be happy."

"Your angel will sing for you and will also smile for you every day. And you will feel your angel's love and be happy."

"And how am I going to be able to understand when people talk to me," the child continued,"if I don't know the language that men talk?"

God patted him on the head and said,"Your angel will tell you the most beautiful and sweet words you will ever hear, and with much patience and care, your angel will teach you how to speak."

第四关：英文经典美文录入（下篇）

"And what am I going to do when I want to talk to you?"

But God had an answer for that question too. "Your angel will place your hands together and will teach you how to pray."

"I've heard that on earth there are bad men, who will protect me?"

"Your angel will defend you even if it means risking her life!"

"But I will always be sad because I will not see you anymore," the child continued warily.

God smiled on the young one. "Your angel will always talk to you about me and will teach you the way for you to come back to me, even though I will always be next to you."

At that moment there was much peace in Heaven, but voices from earth could already be heard. The child knew he had to start on his journey very soon. He asked God one more question, softly, "Oh God, if I am about to leave now, please tell me my angel's name."

God touched the child on the shoulder and answered,"Your angel's name is not hard to remember. You will simply call her Mommy."

藏经阁

盲打指法汇总

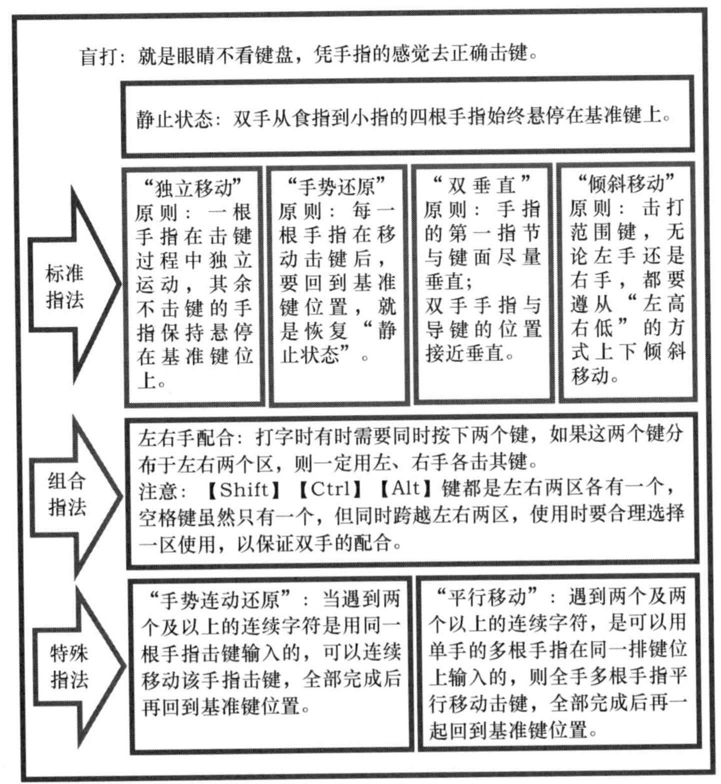

打擂台

以下英文文稿需要录入，记录完成的时间，并参照下表给出自我评价。中级水平的录入速度 250 字符/分钟是每个人通过短期努力都可以达到的。

英文录入自我评价表

级别	完成时间（分钟）	录入速度（字符/分钟）	有效击键速率（次/秒）	称号	文字录入四级鉴定标准
入门	10	150	2.5	打字新手	合格
初级	7.5	200	3.3	打字老手	良好
中、低级	6	250	4.2	打字熟手	优秀
中级	5	300	5	打字高手	
中、高级	4	375	6.25	打字健将	
高级	3	500	8.3	打字达人	

测试文章

文章内容	累计字数
Things do not change; we change. Sell your clothes and keep	60
your thoughts.	74
Think it over.	88
Today we have higher buildings and wider highways, but	143
shorter temperaments and narrower points of view;	192
We spend more, but enjoy less;	222
We have bigger houses, but smaller families;	266
We have more compromises, but less time;	306
We have more knowledge, but less judgment;	348
We have more medicines, but less health;	388
We have multiplied out possessions, but reduced out	440
values;	447
We talk much, we love only a little, and we hate too much;	505
We reached the moon and came back, but we find it	555
troublesome to cross our own street and meet our neighbors;	614
We have conquered the outer space, but not our inner space;	673
We have higher income, but less morals;	712
These are times with more liberty, but less joy;	760
We have much more food, but less nutrition;	803
That's why I propose, that as of today;	842
You do not keep anything for a special occasion, because	898
every day that you live is a special occasion.	944
Search for knowledge, read more, sit on your porch and	999
admire the view without paying attention to your needs;	1054
Spend more time with your family and friends, eat your	1109
favorite foods, visit the places you love;	1151
Life is a chain of moments of enjoyment; not only about	1207
survival;	1216
Remove from your vocabulary phrases like "one of these	1271
days" or "someday";	1290
Let's tell our families and friends how much we love them;	1348
Do not delay anything that adds laughter and joy to your	1405
life;	1410
Every day, every hour, and every minute is special;	1461
And you don't know if it will be your last.	1504

IANGMUER

项目二 中文录入

活动1 改头换面——录入中文

经过一段时间的训练，小文和小璐已经对键盘很熟悉了，英文文章录入速度也达到了一定的水平，他们来到师傅的书房汇报学习情况。

“师傅，师傅，我现在英文录入速度可以达到每分钟180字符啦！”小文兴奋地说。“我也可以打到150个字符！”小璐说。

“口说无凭，我给你们一篇文章，你们拿出自己的真功夫，让我来见识见识！”师傅笑着说道。

……

文章录入完毕，师傅看着两人的成绩，点点头笑眯眯地说：“不错，不错，看来你们都很认真啊！”师傅接着说：“你们还要继续努力练习英文，熟悉键盘，因为我们接下来马上要学习汉字录入了，想要取得好的成绩，指法是关键。”

小文和小璐异口同声回答：“师傅，我们一定会努力练习的，请师傅放心吧！”

“那好，我们就来学习汉字录入吧。”师傅说。

摩拳擦掌

汉字输入的编码方法，基本上都是采用将音、形、义与特定的键相联系，再根据不同汉字进行组合来完成汉字输入的。拼音输入法是按照拼音规定来进行汉字输入的，不需要特殊记忆，符合人的思维习惯，只要会拼音就可以输入汉字。目前主流的拼音输入法是立足于义务教育的拼音知识、汉字知识和普通话水平之上的，所以对使用者普通话、识字及拼音水平的提高有促进作用。我们主要介绍智能ABC输入法和全拼输入法两种。

一、智能ABC输入法

智能ABC输入法（也叫做标准输入法）是运行于Microsoft Windows系统之下的汉语拼音输入法软件，因捆绑于Microsoft Windows简体中文版操作系统而一举成名，曾经是中国大陆使用人数最多的输入法软件。智能ABC输入法状态条如图2—1—1所示。

标准

图2—1—1 智能ABC输入法状态条

二、全拼输入法

全拼输入法就是要将输入汉字的全部拼音录入的一种拼音输入法软件，全拼输入法状态条如图2—1—2所示。

全拼

图2—1—2 全拼输入法状态条

师傅领进门

一、智能 ABC 输入法的使用

智能 ABC 不是一种纯粹的拼音输入法，而是一种音形结合输入法。它不能随着拼音符号的输入立刻显示出字词，而需要先按一下空格键，才能显示出选字菜单，不够直观；而且在输入长句时，句中的每个字、词都要按一下空格加以确认，较为烦琐。这是因为智能 ABC 有一个特殊功能，在音节之后可以输入一个数字，代表某个笔画，这样候选字中就都是以该笔画开头的字，减少（尤其是生僻字的）选字时间。因此在输入拼音的基础上如果再加上该字第一笔形状编码的笔形码，就可以快速检索到这个字。

笔形码所代替的笔形见表 2—1—1。

表 2—1—1　　笔形代码表

代码	1	2	3	4	5	6	7	8
笔形	横	竖	撇	捺、点	左拐	右拐	交叉	方框

例如输入“吴”字，输入“wu8”即可减少检索时翻页的次数，检索范围大大缩小，如图 2—1—3 所示。

二、全拼输入法的使用

输入要打的字的全部拼音字母，例如：“中国”一词，需输入“zhongguo”，如图 2—1—4 所示。

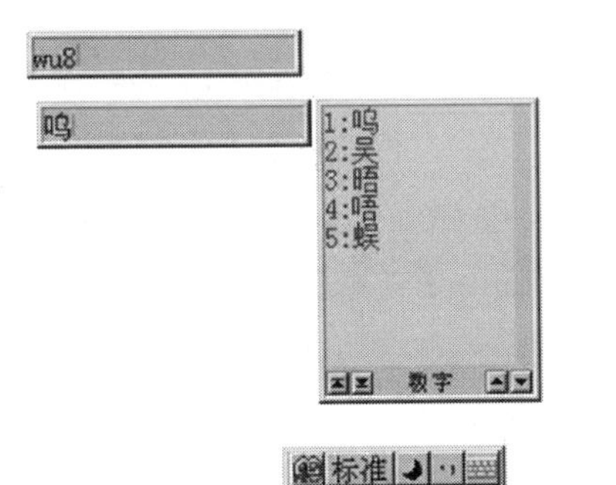

图 2—1—3　智能 ABC 输入法示例

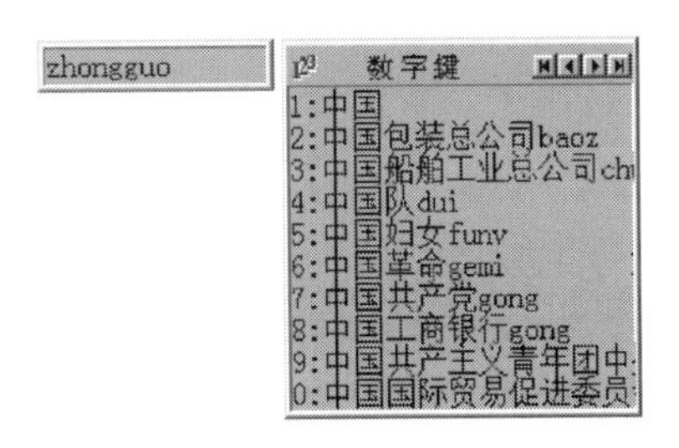

图 2—1—4　全拼输入法示例

修行靠个人

要求：反复练习以下每个部分的输入内容，直到该部分的目标达到以后，才能进入下一个部分的输入练习。

一、全拼输入法录入以下单字

目标：录入速度达到30字符/分钟，即在5分钟以内完成录入。

只以主会样年想能生同老中十从自面前头道它后然走很像
见两用她国动进成回什边作对开而已些现山民候经发工向
事命给长水几义三声于高正妈手知理眼志点心战二问但身
方实吃做叫当住听革打呢真党全才四已所敌之最光产情路
分总条白话东席次亲如被花口放儿常西气五第使写军吧文
关信觉死步反处记将千找争领或师结块跑谁草越字加脚哪

二、智能ABC输入法录入以下单字

目标：录入速度达到30字符/分钟，即在5分钟以内完成录入。

运再果怎定许快明行因别飞外树物活部门无往船望新带队
先力完间却站代员机更九您每风级跟笑啊孩万少直意夜比
阶连车重便斗马哪化太指变社似士者干石满日决百原拿群
究各六本思解立河爸村八难早论吗根共让相研今其书坐接
亮轻讲农古黑告界拉名呀土清阳照办史改历转画造嘴此治
必服雨穿父内识验传业菜爬睡兴形量咱观苦体众通冲合章

灵丹妙药

一、中英文输入切换

在输入中文过程中，需要输入大写英文字母的时候按一下【Caps Lock】键即可，等输完了再按一下【Caps Lock】键就回到中文输入状态。如果想输入小写英文字母，可以先按【v】，然后输入此英文，再按空格键【Space】或回车键【Enter】即可。例如需要在中文输入状态下输入“faint”，直接按【v】faint【空格】就可以了。

二、全半角切换

按【Shift】+【Space】（上档键加空格）即可。例如“～”和“~”，全、半角是不同的。

三、中英文标点符号切换

按【Ctrl】+【.】（控制键和句号）即可。例如“.”和“。”，“￥”和“$”，还有“……”和“^”。

四、简单录入汉字数字

先按【i】，然后再输入想要的数字，按空格即可。例如需输入123456，直接按

【i】【1】【2】【3】【4】【5】【6】【空格】，就可以了。

五、简单录入特殊符号

先按【v】，然后按【1】或者【2】或者【3】，就可以找到很多特殊符号，例如输入【v】【1】，向下翻 7 次，再按【2】，就可以得到“⚲”这个符号；另外，按【v】【2】，可以找到所有的编号排版用符号，如“1.”“（1）”“①”等；按【v】【3】可以找到常见字符的变体，比如“@ ∗{％E”等。

六、简单录入其他字符

先按【v】，然后按【4】，就可以找到所有的日文平假名；按【v】【5】就可以找到日文片假名。例如连续输入【v】【4】【1】，就可以得到“ぁ”。按【v】【6】和【v】【7】可以用来录入希腊字母。按【v】【8】是汉语拼音和我国台湾地区使用的注音符号，比如“ā e ǔ ɑ̄ㄅㄉ”等。按【v】【9】是制图符号，例如“—｜┅┏┓”等。

过关斩将

要求：反复练习以下每部分的输入内容，直到该部分的目标达到以后，才能进入下一部分的测试，全部测试完成后可以进入活动 2。

第一关：单字录入练习

目标：录入速度达到 20 字符/分钟，即在 5 分钟以内完成录入。

的一国在人了有中是年和大业不为发会工经上地市要个产

这出行作生家以成到日民来我部对进多全建他公开们场展

时理新方主企资实学报制政济用同于法高长现本月定化加

动合品重关机分力自外者区能设后就等体下万元社过前面

第二关：文章录入练习

目标：录入速度达到 30 字符/分钟，即在 5 分钟以内完成录入。

船在动，星也在动，它们是这样低，真是摇摇欲坠呢！渐渐地我的眼睛模糊了，我好像看见无数萤火虫在我的周围飞舞。海上的夜是柔和的，是静寂的，是梦幻的。我望着许多认识的星，我仿佛看见它们在对我眨眼，我仿佛听见它们在小声说话。这时我忘记了一切。在星的怀抱中我微笑着，我沉睡着。我觉得自己是一个小孩子，现在睡在母亲的怀里了。

第三关：文章录入练习

目标：录入速度达到 40 字符/分钟，即在 5 分钟以内完成录入。

那是力争上游的一种树，笔直的干，笔直的枝。它的干呢，通常是丈把高，像是加以人工似的，一丈以内，绝无旁枝；它所有的桠枝呢，一律向上，而且紧紧靠拢，也像是加以人工似的，成为一束，绝无横斜逸出；它的宽大的叶子也是片片向上，几乎没有斜生的，更不用说倒垂了；它的皮，光滑而有银色的晕圈，微微泛出淡青色。这是虽在北方的风雪的压迫下却保持着倔强挺立的一种树！哪怕只碗来粗细罢，它却努力向上发展，高到丈许，两丈，参天耸立，不折不挠，对抗着西北风。

活动2 步步为营——录入字根

经过一段时间的汉字录入训练，小文和小璐在机房里交流经验，小文问小璐："这个字怎么拼啊，是前鼻音还是后鼻音啊?"

"哦，'声音'的'声'是后鼻音，而'音'是前鼻音，这个词很容易出错的。"小璐说："我在录入的过程中，不仅这些容易出错的词录入很慢，还有些字要翻页才能找到，有时还会碰到不认识的字。"

"是的，我在录入过程中也会遇到许多问题要停下来，这样录入的速度可真慢呀，我们什么时候才能成为汉字录入高手啊?"小文唉声叹气地说。

"我们还是去问问师傅吧!"

"好的!"

于是两人一起来到书房，并把遇到的问题都说给师傅听。师傅耐心地听完她们的问题，点点头微笑地说："很好，很好，你们不仅认真地训练，而且还从练习中发现了许多问题，这说明你们是用脑子在学习啊!"

师傅接着说："你们遇到的这些问题正是使用拼音输入法录入汉字的缺点所在，因为使用拼音输入法首先是要认识这些字，然后才能录入，另外拼音有许多同音字，因此在录入过程中可能要翻页才能找到，影响到录入速度。为了解决这些问题，我们就要学习五笔字型输入方法，它能解决录入过程中的这些问题。"

摩拳擦掌

五笔字型输入法（简称五笔）是王永民教授发明的一种形码输入法，它是按照汉字的字形（笔画、部首与结构）进行编码的。因为发明人姓王，所以也称为"王码五笔"。86 版五笔字型输入法是目前使用人数最多的一种五笔字型输入法，许多其他类型的五笔字型输入法的编码规则也都是按照 86 版王码五笔字型输入法来开发的。

五笔字型输入法是一种高效率的汉字输入法，它只使用电脑键盘的 25 个字母键，按照汉字的笔画结构，即我们只需把待输入的汉字拆分成已知字根，然后敲击这些字根所在的键，就可将汉字输入进去。它有以下几个特点：重码少、效率高、输入快、字词兼容。

师傅领进门

一、汉字的 3 个层次

汉字有三个层次，在书写汉字时，不间断地一次写成的一个线条叫汉字笔画；由若干笔画复合连接所形成的，相对不变的结构叫字根；将字根按一定的位置关系拼合起来，就构成了单字。由此可见，汉字可以划分为笔画、字根、单字三个层次。

二、汉字的 5 种笔画

如果按其长短、曲直和笔势走向来分，笔画可以分到几十种。为了易于被人们接受和掌握，可对其进行科学的分类。如果按照书写方向来划分，笔画的类型只有 5 种，它们分别是横、竖、撇、捺、折。

● 横（一）

在五笔字型中，“横”是指笔画的走向为从左向右，如“卖”“未”等字的首笔画都属于“横”笔画。另外，提笔在五笔字型中也视为横，如“现”“场”“特”“冲”等字左部的末笔画。

● 竖（丨）

在五笔字型中，“竖”是指笔画的走向为从上到下，如“十”“下”等字中的第二个笔画都属于竖笔画。另外，左竖钩在五笔字型中被视为竖，如“利”“到”等字的最后一笔。

● 撇（丿）

在五笔字型中，“撇”是指笔画的走向为从右上到左下，如“川”字最左边的、“毛”字最上方的和“人”字左侧的“丿”等。

● 捺（㇏）

在五笔字型中，“捺”是指笔画的走向为左上到右下，如“八”“极”等字中的“㇏”均属于“捺”笔画。

另外，“点”在五笔字型中也视为捺（包括“宀”中的点），因为其笔画的走向也为从左上到右下，如“学”“家”“冗”等各字中的点都视为“捺”笔画。

● 折（乙）

在五笔字型中，一切带转折的笔画都归为“折”，如“飞、习、乃”字中的折，右竖钩也为折，如“以、饭”各字中的折等。

在五笔字型输入法中，为了便于记忆和排序，分别用 1，2，3，4，5 作为五种笔画的编码，见表 2—2—1。

表 2—2—1　　汉字的五种笔画

编码	笔画（及其变体）	笔画走向
1	横（提）	左→右
2	竖（左竖钩）	上→下
3	撇	右上→左下
4	捺（点）	左上→右下
5	折（一笔写成的各种带转折笔画，包括右竖钩）	带转折

三、汉字的字根

如前所述，汉字可以看成是由一系列笔画组成的，但这种方法过于烦琐。因此，人们将一些由基本笔画组成的相对不变的结构归纳为所谓的偏旁、部首，如我们平常所说的“木子李”“立早章”等。在五笔字型中，这些偏旁、部首被称为字根。

汉字有很多字根，五笔字型输入法根据其输入方案的需要，精选出 125 种常见的字根，加 5 种笔画字根，共 130 种，把它们分布在电脑的键盘上，作为输入汉字的基本单位。

在五笔字型方案中，字根的选取标准主要基于以下几点：

1. 首先选择那些组字能力强、使用频率高的偏旁部首，如：王、土、大、木、工、目、日、口、田、山、纟、禾、亻等。

2. 组字能力不强、但组成的字在日常汉语文字中出现次数很多，如“白、勺”组成的“的”字可以说是全部汉字中使用频率最高的，因此，“白”被作为基本字根。

3. 绝大多数基本字根都是查字典时的偏旁部首，如：人、口、手、金、木、水、火、土等。相反，相当一部分偏旁部首因为太不常用，或者可以拆成几个字根，便不作为基本字根，如：比、风、气、足、老、业、斗、酉、骨、殳、欠、麦等。

4. 五笔字型常用的基本字根是 125 种。有时，一个字根中还包含有几个“小兄弟”，它们主要是：

- 字源相同的字根：心、忄；水、氺、氵、水、氵等。
- 形态相近的字根：艹、廾、廿；已、己、巳等。
- 便于联想的字根：耳、卩、阝等。

四、字根的键盘布局、字根总表和助记词

在五笔字型中，将 125 个常用的基本字根按起笔的类型（横、竖、撇、捺、折）分为 5 类，每一类又分为 5 组，共计 25 组。同时，将键盘上除【Z】键以外的 25 个字母划分为五个区，将这 5 类字根分别放置在 5 个区中，每一类的 5 个组又分别与每一

区中的 5 个键位相对应，如图 2—2—1 所示。

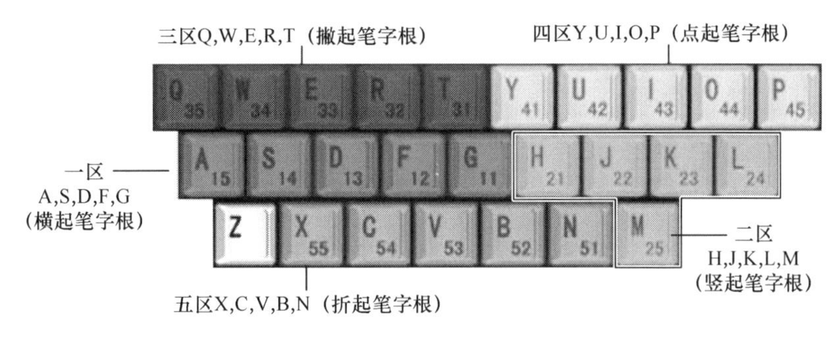

图 2—2—1　键盘分区图

其中，区号和位号的定义原则如下：

1. 区号按起笔的笔画横、竖、撇、捺、折划分。如禾、白、月、人、金的首笔均为撇，撇的代号为 3，故它们都在 3 区。也可以说，以撇为首笔的字根，其区号为 3。

2. 一般来说，字根的次笔代号尽量与其所在的位号一致，如土、白、门的第 2 笔均为竖，竖的代号为 2，故它们的位号都为 2。但并非完全如此，例如“工”字的次笔为竖（代号应为 2），但它却被放在了 15 位，而不是 12 位。

3. 复笔画字根的数值尽量与位号一致。例如，单笔画一、丨、丿、㇏、乙都在第 1 位，两个单笔画的复合字根二、‖、⺀、冫、巜都在第 2 位，3 个单笔画的复合字根三、|||、彡、氵、巛都在第 3 位，依此类推。

总之，任何一个字根都可以用它所在的区位号（也叫字根的“代码”）来表示。如“目”字在 2 区 1 位，其区位号为 21，21 就是“目”的代码。五笔字型的字根总图如图 2—2—2 所示。

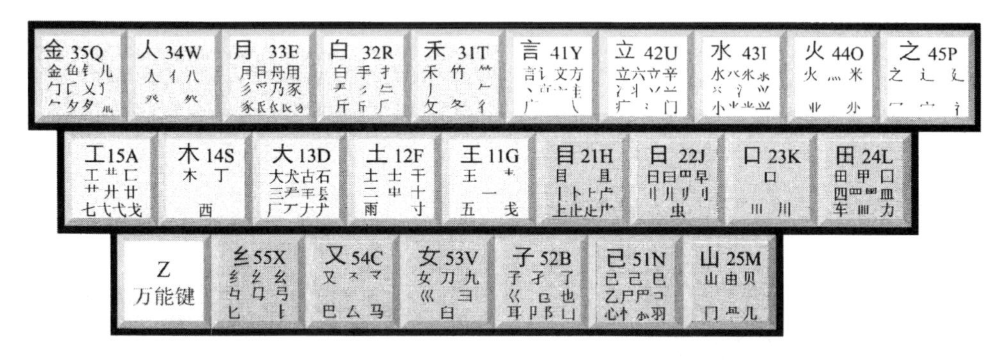

图 2—2—2　五笔字型的字根总图

五笔字型的字根总表及助记词见表 2—2—2。

表 2—2—2　　五笔字型的字根总表及助记词

区	位	代码	字母	键名	基本字根	助记词
1 横起笔类	1	11	G	王	龶 戋 五 一	王旁青头戋（兼）五一
	2	12	F	土	士 二 干 十 寸 雨	土士二干十寸雨
	3	13	D	大	犬 三 𡗗 古 石 厂 丆 ナ 镸	大犬三（羊）古石厂
	4	14	S	木	丁 西	木丁西
	5	15	A	工	戈 弋 艹 艹 廾 廿 匚 七	工戈草头右框七
2 竖起笔类	1	21	H	目	丨 且 上 ⺊ 止 龰 卜 虍 ⺶	目具上止卜虎皮
	2	22	J	日	曰 早 刂 刂 川 刂 虫 罒 虫	日早两竖与虫依
	3	23	K	口	川 川	口与川，字根稀
	4	24	L	田	甲 囗 四 皿 车 力	田甲方框四车力
	5	25	M	山	由 贝 冂 几 冎	山由贝，下框几
3 撇起笔类	1	31	T	禾	竹 ⺮ 𠂉 丿 彳 攵 夂	禾竹一撇双人立 反文条头共三一
	2	32	R	白	手 扌 ⺧ 𡗗 厂 斤	白手看头三二斤
	3	33	E	月	彡 乃 用 豕 爫 丹 乡 豕 𧘇 ⺽	月衫乃用家衣底
	4	34	W	人	亻 八 癶	人和八，三四里
	5	35	Q	金	钅 勹 ⺈ 鱼 犭 儿 乂 儿 夕 夕 𠂆	金勺缺点无尾鱼 犬旁留乂儿一点夕 氏无七
4 捺起笔类	1	41	Y	言	讠 文 方 广 亠 丶 圭	言文方广在四一 高头一捺圭（谁）人去
	2	42	U	立	辛 丷 䒑 冫 丬 六 立 门 疒	立辛两点六门疒（病）
	3	43	I	水	氵 氺 水 氵 ⺍ 𭕄 小 ⺌ ⺍	水旁兴头小倒立
	4	44	O	火	⺗ 业 灬 米	火业头，四点米
	5	45	P	之	廴 辶 冖 宀 礻	之宝盖，摘礻（示）衤（衣）
5 折起笔类	1	51	N	已	巳 己 コ 乙 尸 心 忄 ⺗ 羽	已半巳满不出己 左框折尸心和羽
	2	52	B	子	孑 阝 耳 卩 㔾 了 也 凵 巜	子耳了也框向上
	3	53	V	女	刀 九 臼 彐 巛	女刀九臼山朝西
	4	54	C	又	巴 马 厶 マ 龴	又巴马，丢矢矣
	5	55	X	纟	弓 匕 纟 幺 幺 ⺕	慈母无心弓和匕，幼无力

五、巧记各区字根

1. 一区：如图 2—2—3 所示。

图 2—2—3　一区字根

【G】键（区位码 11)：王旁青头戋（兼）五一。“青头”指“青”的上半部分，即“龶”；“戋”与“兼”同音，“五”与“王”形近。

【F】键（区位码 12)：土士二干十寸雨。“士”与“土”同形；“干”为倒“士”；“卑”与“十”形似。

【D】键（区位码 13)：大犬三𡗗（羊）古石厂。“𡗗”指羊字底“𡗗”和“龵”；“丆”和“ナ”与“厂”形似；“犬”与“尢”相似。

【S】键（区位码 14)：木丁西。该键位上只有 3 个字根。

【A】键（区位码 15)：工戈草头右框七。“戈”指的是字根“弋、戈”；“草头”指的是“草”字的上部即“艹”及其变形体“廾、廿、龷”；“右框”（即框向右而不是向左）指的是“匚”。

2. 二区：如图 2—2—4 所示。

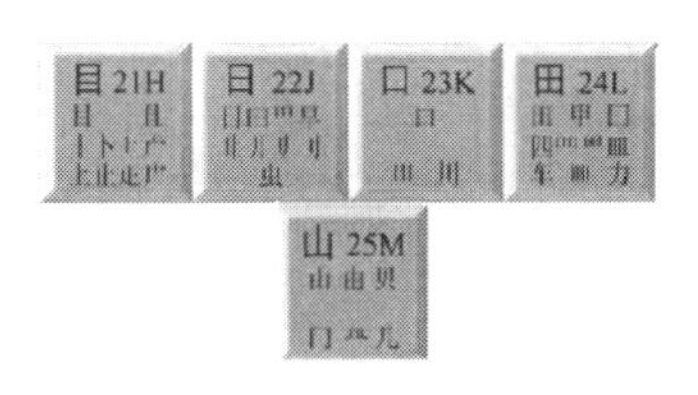

图 2—2—4　二区字根

【H】键（区位码 21)：目具上止卜虎皮。“具上”指的是“具”字的上半部分“且”；“龰”为“止”的变形体；“⺊”为“卜”的变形体；“虎皮”指的是“虎”、“皮”两字的上部（虍、⺁）。

【J】键（区位码 22)：日早两竖与虫依。“日”是指字根“日、曰、⺫”；“两竖”是指字根“丨丨、丿丨、刂、刂”（其中包含一些变形体)；“与虫依”指的是跟“虫”在同一个键上。

【K】键（区位码 23)：口与川，字根稀。“川”指的是“川、丨丨丨”字根；“字根稀”是指该键上的字根不多。

【L】键（区位码 24）：田甲方框四车力。“方框”指的是“口（不是‘口与川’中的‘口’）”；“四”指的是“四、罒、皿、罒”字根。

【M】键（区位码 25）：山由贝，下框几。“下框”指方框向下即“冂、几”字根。

3．三区：如图 2—2—5 所示。

图 2—2—5 三区字根

【T】键（区位码 31）：禾竹一撇双人立，反文条头共三一。“禾”指的是“禾、禾”字根；“竹”指的是“竹、⺮”字根；“一撇”指的是“丿、𠂉”字根；“双人立”指的是“彳”；“反文”指的是“攵”；“条头”指的是“条”字的上部“夂”；“共三一”指的是它们都在区位码为 31 的键位上。

【R】键（区位码 32）：白手看头三二斤。“手”指的是“手、扌”字根；“看头”指的是“看”的上部“𠂇”；“三二”指的是这些字根都在区位码为 32 的键位上；“斤”指的是“斤、斤”字根。

【E】键（区位码 33）：月彡（衫）乃用家衣底。“月”指的是“月、⺼、丹”字根；“彡”指三撇及“爫”字根；“家衣底”指的是这三个字的下半部分及其变形体即“豕、豸、𧘇、𠄌”。

【W】键（区位码 34）：人和八，三四里。“人”还包括字根“亻”；“八”指的是“八”及其变形体“癶”；“三四里”指它们都在区位码为 34 的键位上。

【Q】键（区位码 35）：金勺缺点无尾鱼，犬旁留乂儿点夕，氏无七。“金”指的是“金、钅”字根；“勺缺点”指的“勺”字缺一点即“勹”；“无尾鱼”指“鱼”没有尾巴即“鱼”；“犬旁”指的是“犭”字根；“留乂”指的是“乂”字根；“儿”与“儿”相似；“点夕”指带一点的“夕”、少一点的“ク”、多一点的“夕”；“氏无七”指“氏”字去掉“七”即“𠂆”。

4．四区：如图 2—2—6 所示。

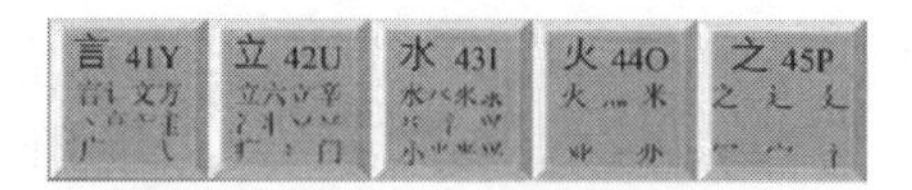

图 2—2—6 四区字根

【Y】键（区位码 41）：言文方广在四一，高头一捺谁人去。“言”还包括“讠”字根；“在四一”指的是它们在区位码为 41 的键位上；“高头”指的是“高”的上部“亠、古”；“一捺”是指“㇏”和“丶”；“谁人去”指的是“谁”字去掉“亻”和“讠”即留下“主”。

【U】键（区位码 42）：立辛两点六门疒（病）。“六”还包括“六”字根；“两点”指的是“冫、丷、䒑、丬”字根。

【I】键（区位码 43）：水旁兴头小倒立。“水旁”指的是“水、氵、氺、水、氵”字根；“兴头”是指“⺍、⺌、⺍”；“小倒立”是指“小、⺌”。

【O】键（区位码 44）：火业头，四点米。“业头”是指“业”的上部及其变形体“⺌、⺌”；“四点”是指“灬”字根。

【P】键（区位码 45）：之宝盖，摘礻（示）衤（衣）。“之”是指“之”及其变形体“辶、廴”；“宝盖”是指“宀、冖”字根；“摘礻（示）衤（衣）”是指把“礻”“衤”偏旁下方的一点或两点去掉，即为“礻”字根。

5. 五区：如图 2—2—7 所示。

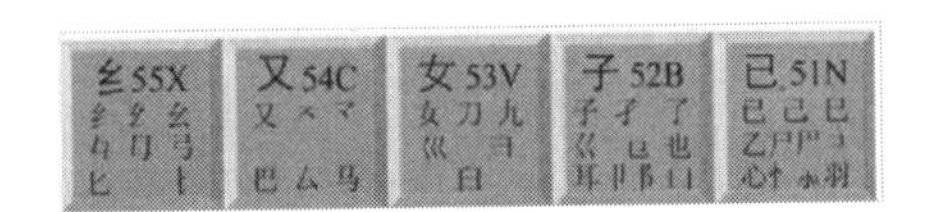

图 2—2—7 五区字根

【N】键（区位码 51）：已半巳满不出己，左框折尸心和羽。“已半”是指半封口的“已”；“巳满”是指全封口的“巳”；“不出己”是指不封口的“己”；“左框”是指开口向左的方框“コ”；“折”是指“乙”；“尸”是指“尸”及其变形体“尸”；“心”是指“心”及其变体“忄、⺗”。

【B】键（区位码 52）：子耳了也框向上。“子”是指“子、孑”；“耳”是指“耳、阝、卩、㔾”字根；“框向上”是指“凵”字根；折笔画数为 2 的“巛”字根也在该键上。

【V】键（区位码 53）：女刀九臼山朝西。“山朝西”是指方向朝西的山即“彐”；折笔画数为 3 的“巛”字根也在该键上。

【C】键（区位码 54）：又巴马，丢矢矣。“又”还包括其变体“ス、マ”；“丢矢矣”是指“矣”字去掉“矢”即“厶”。

【X】键（区位码 55）：慈母无心弓和匕，幼无力。“慈母无心”是指“母”字去掉中间部分即“口”字根；“幼无力”是指把“幼”字去掉“力”即为“幺”，另外还有其变体“乡”。

修行靠个人

要求：在上机练习前先写出下列字根的编码，然后通过上机练习加以验证，再通过反复的上机练习（10～20遍）达到巩固和熟练掌握的目的。

一、填写字根所在的区位号与编码

一区：

西（　）（　）弋（　）（　）士（　）（　）厂（　）（　）木（　）（　）

二（　）（　）丁（　）（　）艹（　）（　）五（　）（　）雨（　）（　）

戋（　）（　）犬（　）（　）廿（　）（　）干（　）（　）古（　）（　）

寸（　）（　）镸（　）（　）七（　）（　）龶（　）（　）廾（　）（　）

二区：

刂（　）（　）卜（　）（　）皿（　）（　）冂（　）（　）川（　）（　）

口（　）（　）罒（　）（　）上（　）（　）亅（　）（　）力（　）（　）

虫（　）（　）止（　）（　）曰（　）（　）几（　）（　）目（　）（　）

山（　）（　）甲（　）（　）贝（　）（　）车（　）（　）田（　）（　）

三区：

乃（　）（　）龵（　）（　）彡（　）（　）亻（　）（　）攵（　）（　）

⺮（　）（　）𠂉（　）（　）扌（　）（　）爫（　）（　）夂（　）（　）

豕（　）（　）禾（　）（　）癶（　）（　）钅（　）（　）彳（　）（　）

厂（　）（　）八（　）（　）月（　）（　）用（　）（　）白（　）（　）

四区：

辶（　）（　）立（　）（　）方（　）（　）文（　）（　）灬（　）（　）

六（　）（　）冖（　）（　）讠（　）（　）疒（　）（　）⺍（　）（　）

之（　）（　）辛（　）（　）门（　）（　）亠（　）（　）氵（　）（　）

丬（　）（　）水（　）（　）言（　）（　）丷（　）（　）米（　）（　）

五区：

已（　）（　）女（　）（　）尸（　）（　）耳（　）（　）㔾（　）（　）

子（　）（　）幺（　）（　）心（　）（　）彐（　）（　）马（　）（　）

臼（　）（　）忄（　）（　）凵（　）（　）巴（　）（　）乙（　）（　）

刀（　）（　）弓（　）（　）也（　）（　）羽（　）（　）纟（　）（　）

二、填写基本字根的编码

丬（　　　）广（　　　）口（　　　）纟（　　　）山（　　　）

七（ ）龶（ ）厂（ ）八（ ）丿（ ）
月（ ）镸（ ）刀（ ）匕（ ）廾（ ）
白（ ）匚（ ）羽（ ）⺌（ ）寸（ ）
镸（ ）车（ ）由（ ）甲（ ）贝（ ）
斤（ ）弓（ ）用（ ）丷（ ）米（ ）
言（ ）水（ ）火（ ）也（ ）𠂇（ ）
西（ ）豕（ ）臼（ ）虫（ ）癶（ ）
皿（ ）止（ ）爫（ ）立（ ）雨（ ）
子（ ）戋（ ）辛（ ）幺（ ）攵（ ）

三、将汉字拆分成基本字根

列（ ）责（ ）示（ ）坛（ ）无（ ）
碌（ ）载（ ）应（ ）较（ ）折（ ）
罗（ ）互（ ）家（ ）比（ ）波（ ）
如（ ）底（ ）销（ ）符（ ）离（ ）
陷（ ）展（ ）刘（ ）孜（ ）框（ ）
宽（ ）请（ ）社（ ）分（ ）失（ ）
断（ ）彩（ ）建（ ）设（ ）好（ ）
型（ ）炉（ ）冰（ ）曲（ ）架（ ）
防（ ）敲（ ）谈（ ）鲜（ ）因（ ）
农（ ）凡（ ）宝（ ）远（ ）陈（ ）

灵丹妙药

一、五种笔画的判断

五种笔画的变形体不拘一格：竖带左钩的笔画视为竖，如“丁”“小”中的竖钩；提笔因是由左向右视为横，如地、扣、刁的提笔；点视为捺，如六、立、注、兴中的点；竖笔向右钩和其他只要代拐笔画均视为折，如氏、匕、以中的折笔。因此在判断笔画的种类时，要特别注意根据笔画的书写方向来判断。另外，五种单笔画横、竖、撇、点、捺也是基本字根。

二、键位排列

五笔字型字根在键盘上的分布大部分按以下原则。

1. 按第一笔的笔画分区：根据前面所说的五种笔画的代码，按字根第一笔的代码确定该字根在哪一区。也就是说把首笔是横、竖、撇、捺、折的分别分到 12345 区。

2. 按第二笔的笔画定位：把第二笔笔画代码安排该字根在几号位。如“王”字，第一笔为横，可知区代码为 1，第二笔仍为横，位码也是 1，于是，安排在 11 键上；如“士”字，首笔为横，第二笔为竖，区码 1，位码 2，就在 12 键上；如“七”字，首笔为横，第二笔为折，就在 15 键上；如“之”字，首笔为点，第二笔为折，就在 45 键上。

3. 按笔画数定位：如横笔一横、二横、三横，分别在 11，12，13 键；一竖、二竖、三竖、四竖分别在 21，22，23，24 键；一撇、二撇、三撇分别在 31，32，33 键；一点、二点、三点、四点分别在 41，42，43，44 键；一折、二折、三折分别在 51，52，53 键。

4. 按以上规则分配后，有些位上分配字根较少，可将字根分布过于集中的键上的字根调剂过去。如汉字书写笔画中没有首笔为横或竖，第二笔为捺的字根，于是在 14 和 24 键上分别安排了“木、丁、西”和“田、甲、车”，这都是从其他键位调剂过来的。

5. 按汉字传统偏旁部首有相应关系的，虽笔画走向不同，为便于记忆，也安排在一起，如水、耳等。

总之，五笔字型方案的键位排列，既考虑了各个键位的使用频率和键盘指法，又使字根代号从键盘中央向两侧依大小顺序排列。这样做的好处是键位便于掌握、代号好学好记、操作员易于培训、击键效率便于提高。

三、五笔助记词

为了便于学习与掌握，五笔字型的发明者对每一个区的字根都编写了助记口诀，不但押韵上口，还有些“诗味”，多念几遍，便能记住各键位有哪些字根。

过关斩将

要求：反复练习以下每部分的输入内容，直到该部分的目标达到后，才可进入下一部分的测试，全部测试完成后可进入活动 3。

第一关：一区、二区字根录入练习

目标：录入速度达到 30 字根/分钟，即在 5 分钟以内完成录入。

廿十上卢二丁十十廾犬刂罒ナ二口古二皿干二扌耂二西二工丁匚车弋
木贝早罒川石龷石冂一龶卢一刂目犬ナ艹川刂三田大卜戈川戋皿弋广
广石四扌上串皿五四口皿亅刂皿车丁龷虫力石十亅止日上艹甲刂丆龷
丆几士寸七甲龷日士止扌五丁镸贝干日刂龷丨七卢犬力甲由艹由止石
川十川卜雨干囗干工龶刂上扌扌曰囗一皿二罒镸且田几四甲廿早刂冂

第二关：三区、四区、五区字根录入练习

目标：录入速度达到 30 字根/分钟，即在 5 分钟以内完成录入。

丷门㐅月夕乙儿纟手巳丷冖竹卩巳立耳灬巳巴丬宀亠乀巳斤氺丬夂六
厂竹鱼白亠冖手儿灬乂丷疒臼冖月冫亻⺢月廴尸攵丷亅纟龶氵水彐⺍
卩コ⺧纟臼⺌勹冖也马乀金巛立卩人㐅臼丷女阝鱼彡广鱼⺢彡幺龶辛
宀巳氺人辛尸亠鱼㐅彐氺文⺌⺌夕㐅礻冫⺌㐅冫文广冫尸㐅巳又疒金
纟水川亅㔾心斤斤彳亠疒丷又彐己⺌夕幺儿丷月门亻文礻弓丷豕了斤

第三关：全字根录入练习

目标：录入速度达到 30 字根/分钟，即在 5 分钟以内完成录入。

乙匕凵月二火月丷龶竹言辶几水止虍厶手儿龶弋羽立雨氺⺌丷也乃亅
コ羽丰丹豕虍乀マ弓冫丶女彐月耳亠𠂇豕儿廾月丁门亠小纟丹廾卩车
犬𠂇禾彳丷疒纟人⺌攵𠂇口且𠂉八目日六钅手厶子丷镸辛灬罒虫𠂇𠂇
𠂉雨刀目𠂉尸弋乂虍卩卜亻尸马八又廾厂亠曰曰巳𠂉乂由丷士巴皿米
廴几卩八三宀八广二コ𠂉卜纟夂钅子田纟⺢文止用月九月子亠廿九止

活动 3　名不虚传——录入键名汉字

经过刻苦地练习，小文和小璐已经把字根助记词和字根表记得滚瓜烂熟了，两人也互相进行了几次较量，成绩不相上下。小文提议说："我们去找师傅吧，让他来考考我们，看看谁练得好。"

"好呀，我不会输给你的。走，让师傅来评评！"小璐点头同意。

他们来到师傅的书房，小文抢先说："师傅，我已经把字根助记词背得很熟了，王旁青头兼五一，土士二干十寸雨……"

"师傅，我也背出来了，你听……"小璐也不甘示弱。

师傅哈哈大笑并赶紧说："好好好，我知道你们都很认真，那我就来考考你们，请你们坐到计算机前，测试正式开始。"

……

测试完毕，师傅看看测试速度满意地对小文和小璐说："嗯，你们俩的成绩都不错，达到了要求，可以继续学习五笔录入。""那么，你们还记得汉字的三个层次吗？"师傅接着问。

“笔画、字根、单字。”小璐抢答道。

“对，对，那你们看看这三个层次，就应该知道我们接下去要学习什么内容了吧！”

“应该是单字吧？”小文轻声地回答。

“对了，这就是今天我们要学习的内容，单字录入的第一步：键名汉字录入。”

摩拳擦掌

按照五笔字型输入法的规定，130 个字根依据“横、竖、撇、捺、折”的原则，按 5 个区 5 个位有序地分布在 A—Y 的 25 个键位上，每个键上取一个字根作为键名，其名谱如下。

一区：横起笔，王 土 大 木 工

二区：竖起笔，目 日 口 田 山

三区：撇起笔，禾 白 月 人 金

四区：捺起笔，言 立 水 火 之

五区：折起笔，已 子 女 又 纟

要记住各个键位上的字根，应借助字根助记词，每句字根助记词的第一个字就是键名汉字。熟记键名汉字的最好方法就是背熟键位表，五笔字型的键位表及键名汉字如图 2—3—1 所示。

图 2—3—1　键位表及键名汉字

师傅领进门

五笔字型输入法规定，键名汉字的编码由它所在的键的代码重复 4 次组成（助记词——键名汉字连击四下）。这就是说，键名汉字的编码中，4 个代码都是一样的，都是它所在键的代码。例如：

王：11 11 11 11 (GGGG)

木：14 14 14 14 (SSSS)

口：23 23 23 23 (KKKK)

根据这个规定，输入键名汉字时，就把键名汉字所在键连击四下就可以了。

修行靠个人

要求：在上机练习前先写出下列键名汉字的编码，然后通过上机练习加以验证，再通过反复的上机练习（10～20 遍）达到巩固和熟练掌握的目的。

大（　　　）田（　　　）之（　　　）禾（　　　）纟（　　　）
子（　　　）女（　　　）山（　　　）王（　　　）土（　　　）
木（　　　）工（　　　）目（　　　）金（　　　）已（　　　）
月（　　　）日（　　　）口（　　　）言（　　　）又（　　　）
水（　　　）人（　　　）火（　　　）立（　　　）白（　　　）

灵丹妙药

特殊键名汉字：有些键名汉字不必连击所在键四下，如“人”和“水”等。这是由于可以使用简码输入汉字，关于这部分内容，将在简码中介绍。

过关斩将

要求：反复练习以下每部分的输入内容，直到该部分的目标达到后，才可进入下一部分的测试，全部测试完成后可以进入活动 4。

第一关：一、二区键名汉字录入练习

目标：录入速度达到 30 字/分钟，即在 2 分钟以内完成录入。

田 山 王 大 田 山 王 土 木 工 目 日 口 大 木 工 目 日 土 口
大 木 工 目 王 土 大 木 田 山 土 大 木 工 目 口 工 目 王 土
王 土 大 木 田 山 工 目 日 口 大 工 目 口 木 田 山 大 土 王

第二关：三、四、五区键名汉字录入练习

目标：录入速度达到 30 字/分钟，即在 2 分钟以内完成录入。

禾 白 又 纟 金 言 立 水 火 之 白 月 人 立 水 禾 言 立 月 人
言 金 子 女 已 月 水 人 白 月 火 人 女 禾 之 言 立 白 月 水
之 禾 纟 子 女 又 人 火 人 之 禾 白 月 女 又 立 禾 月 水 人

第三关：键名汉字录入练习

目标：录入速度达到 30 字/分钟，即在 2 分钟以内完成录入。

大 田 之 禾 纟 子 女 山 王 土 木 工 目 金 子 女 已 月 工 目

日 口 之 言 金 子 女 已 月 水 纟 土 大 木 田 山 人 火 人 之
立 水 土 工 目 纟 口 田 山 禾 白 大 木 月 金 火 之 已 子 又

活动 4 一笔一画——录入成字字根

小文和小璐按照约定的时间来到了师傅的书房。

“你们俩练得怎么样了？”师傅笑着问道。

“师傅，我字根已经牢记在心，键名汉字也已经烂熟于心了。”小璐抢先回答，同时得意地瞟了一眼身旁的小文。“嗯，那小文你呢？”师傅转向小文问道。“耳听为虚，眼见为实，要不师傅你就当场考考我们俩吧。”小文说完，回赠给小璐一个自信的眼神。

“好，要学习今天新知识的前提就是你们必须能够熟练地打出 130 个基本字根和 24 个键名汉字（另有 1 个键名字根），”师傅肯定地回复道，“所以，我必须了解你们现在的训练情况，那现在就请你们各就各位，测试正式开始。”

……

“嗯，不错不错，你们俩的成绩都达到了我的要求，看来你们回去都下工夫练习了。”师傅边看着俩人的测试结果边频频点头。话锋一转，师傅接着说道：“在 130 个基本字根里，除了键名汉字以外，还有其他一些本身也是汉字的字根，比如……”

不等师傅讲完，小文和小璐就争先恐后你一个我一个地抢答起来，“比如：马、手、刀、雨、贝、古、川、……。”师傅连忙打住他们说：“好啦好啦，师傅知道你们对于基本字根已经掌握得滚瓜烂熟了，那你们知道这些汉字字根叫做什么吗？它们又该如何录入呢？”

小文和小璐茫然地对望着摇了摇头，不约而同地回答：“不知道。”师傅见状笑着说道：“这就是我今天要教给你们的新知识：录入成字字根。”

摩拳擦掌

在开始学习成字字根之前，让我们先来找一找，看看在字根总表的每个键位上除了键名汉字以外，还有哪些汉字？现归纳总结见表 2—4—1。

表 2—4—1　　　　**字根总表中除键名汉字以外的汉字**

分区	键位	除键名汉字以外的汉字
1 区	G	戋、五、一
	F	士、二、干、十、寸、雨
	D	犬、三、古、石、厂
	S	丁、西
	A	戈、弋、廿、七
2 区	H	上、止、卜
	J	曰、早、虫
	K	川
	L	甲、四、皿、车、力
	M	由、贝、几
3 区	T	竹
	R	手、斤
	E	乃、用、豕
	W	八
	Q	儿、夕
4 区	Y	文、方、广
	U	辛、六、门
	I	小
	O	米
5 区	N	巳、己、乙、尸、心、羽
	B	耳、了、也、子
	V	刀、九、臼
	C	巴、马
	X	弓、匕、幺

从上表可以看出，除了键名汉字外，还有 70 个其他汉字，进一步观察还可以发现，在这 70 个汉字中包含了两个单笔画汉字：一、乙。

你们还记得“横、竖、撇、捺、折”这五个单笔画所对应的键位吗？让我们复习一下吧。

横（一）：G

竖（丨）：H

撇（丿）：T

捺（丶）：Y

折（乙）：N

师傅领进门

在 130 个字根中，除 24 个键名汉字外，还有一些本身就是汉字的字根，被称为成字字根。

成字字根共有 68 个，见表 2—4—1 中除了“一”和“乙”之外的所有汉字；“一”和“乙”称为单笔画汉字，它们和“丨”“丿”“丶”一起构成了单笔画。成字字根和单笔画的编码规则是不同的。

一、成字字根的编码

成字字根的编码公式为：

报户口＋首笔笔画＋次笔笔画＋末笔笔画

注意：

1.“报户口”就是敲击成字字根所在键字母。

2. 首笔笔画、次笔笔画和末笔笔画不是按字根取码，而是按构成成字字根笔画顺序的各个单笔画取码，即横、竖、撇、捺、折这五个单笔画所分别对应的 G，H，T，Y，N 键取码。

3. 如果成字字根仅有两笔笔画，则末笔笔画用空格代替。

示例见表 2—4—2。

表 2—4—2　成字字根编码示例

例字	报户口	首笔笔画	次笔笔画	末笔笔画
由	M	H	N	G
马	C	N	N	G
手	R	T	G	H
九	V	T	N	空格
刀	V	N	T	空格

二、单笔画的编码

“一”和“乙”虽然是作为汉字的字根，但是由于它们只有一个笔画，因此不适用于成字字根的编码公式，而单笔画有时也需要单独使用，因此特别规定五个单笔画的编码公式为：

报户口＋首笔笔画＋LL

示例

一：GGLL

丨：HHLL

丿：TTLL

丶：YYLL

乙：NNLL

24 个键名汉字、68 个成字字根和 2 个单笔画汉字合称键面字，共计 94 个。

修行靠个人

要求：在上机练习前先写出下列汉字的编码，然后通过上机练习加以验证，之后再通过反复的上机练习（10～20 遍）达到巩固和熟练掌握成字字根和单笔画汉字的目的。

一、一区的成字字根（共 19 个）

戋（　　）五（　　）士（　　）二（　　）

干（　　）十（　　）寸（　　）雨（　　）

犬（　　）三（　　）古（　　）石（　　）

厂（　　）丁（　　）西（　　）戈（　　）

弋（　　）廿（　　）七（　　）

二、二区的成字字根（共 15 个）

上（　　）止（　　）卜（　　）曰（　　）

早（　　）虫（　　）川（　　）甲（　　）

四（　　）皿（　　）车（　　）力（　　）

由（　　）贝（　　）几（　　）

三、三区的成字字根（共 9 个）

竹（　　）手（　　）斤（　　）乃（　　）

用（　　）豕（　　）八（　　）儿（　　）

夕（　　）

四、四区的成字字根（共 8 个）

文（　　）方（　　）广（　　）辛（　　）

六（　　）门（　　）小（　　）米（　　）

五、五区的成字字根（共 17 个）

已（ ）己（ ）尸（ ）心（ ）
羽（ ）耳（ ）了（ ）也（ ）
子（ ）刀（ ）九（ ）臼（ ）
巴（ ）马（ ）弓（ ）匕（ ）
幺（ ）

六、单笔画汉字（共 2 个）

一（ ）乙（ ）

灵丹妙药

一、圈圈点点

从报纸或杂志中任意找一篇文章，把上面的成字字根及单笔画汉字圈出来，然后逐一校对检查。如此反复多次，可以提高自己对于成字字根和单笔画汉字的敏感度，最终达到能“下意识”地进行输入的程度。

二、注意笔顺

成字字根是根据构成该成字字根的各个单笔画的顺序来进行编码的，因此对于每一个成字字根的书写顺序一定要特别清楚，对于自己平时没有按照规范笔顺进行书写的成字字根要特别留意，对其编码要重点记忆。

三、科学练习

在使用打字软件练习时，初期可以使用键盘帮助，在基本入门之后就要关闭键盘帮助。在练习时，要注意进一步理解成字字根和单笔画的编码规则，以达到巩固和熟练掌握的目的。

四、巧用简码

五笔字型编码的标准码长为 4，为了简化输入，提高录入速度，设计者设计了简码输入法，将简码分为一、二、三级，分别只需先敲击相应汉字的前一、二、三个字母键再击一个空格键，即可输入其对应的汉字。关于简码的详细内容可以参见本项目活动 8 和活动 9。

成字字根中有一些汉字就属于简码字，在上机练习过程中，可以对这些汉字进行简码输入。从一开始就养成简码输入的好习惯，有助于今后对简码的熟练应用，从而提高输入效率。本项目的藏经阁部分列出了成字字根的编码及其简码。

过关斩将

要求：反复进行以下每部分的内容测试练习，直到达到该部分的目标后，才能进入下一部分的测试，全部测试完成后可以进入活动 5。

第一关：成字字根（含单笔画汉字）按区位顺序录入

目标：在 10 分钟内完成，即达到 30 字/分钟。

戋五一士二干十寸雨犬三古石厂丁西戈弋廿七上止卜曰早虫川甲四皿车力由贝几竹手斤乃用豕八儿夕文方广辛六门小米巳己乙尸心羽耳了也子刀九臼巴马弓匕幺戋五一士二干十寸雨犬三古石厂丁西戈弋廿七上止卜曰早虫川甲四皿车力由贝几竹手斤乃用豕八儿夕文方广辛六门小米巳己乙尸心羽耳了也子刀九臼巴马弓匕幺戋五一士二干十寸雨犬三古石厂丁西戈弋廿七上止卜曰早虫川甲四皿车力由贝几竹手斤乃用豕八儿夕文方广辛六门小米巳己乙尸心羽耳了也子刀九臼巴马弓匕幺戋五一士二干十寸雨犬三古石厂丁西戈弋廿七上止卜曰早虫川甲四皿车力由贝几竹手斤乃用豕八儿夕文方广辛六门小米巳己乙尸心羽耳了也子刀九臼巴马弓匕幺

第二关：成字字根（含单笔画汉字）乱序录入

目标：在 10 分钟内完成，即达到 30 字/分钟。

戋竹幺手五斤匕几一乃弓贝士用上马由二豕巴力干八臼车十儿九皿寸四刀夕雨甲犬文子川三也方古虫了广早石辛耳曰厂六羽门丁卜心小西止尸米戈七乙巳弋廿己戋竹幺手五斤匕几一乃弓贝士用上马由二豕巴力干八臼车十儿九皿寸四刀夕雨甲犬文子川三也方古虫了广早石辛耳曰厂六羽门丁卜心小西止尸米戈七乙巳弋廿己戋竹幺手五斤匕几一乃弓贝士用上马由二豕巴力干八臼车十儿九皿寸四刀夕雨甲犬文子川三也方古虫了广早石辛耳曰厂六羽门丁卜心小西止尸米戈七乙巳弋廿己戋竹幺手五斤匕几一乃弓贝士用上马由二豕巴力干八臼车十儿九皿寸四刀夕雨甲犬文子川三也方古虫了广早石辛耳曰厂六羽门丁卜心小西止尸米戈七乙巳弋廿己

第三关：键面字（键名汉字、成字字根、单笔画汉字）混合录入

目标：在 12 分钟内完成，即达到 30 字/分钟。

目戋己又廿竹日幺弋手巳口乙五金斤上七匕戈禾米几立一尸乃止火西弓人贝小士心王卜用言马丁由门之羽二工豕大六巴厂山曰力土耳干辛八石子已臼早车广十了白水儿虫九古皿方木月寸也四三刀川女田夕雨子甲文犬目戋己又廿竹日幺弋手巳口乙五金斤上七匕戈禾米几立一尸乃止火西弓人贝小士心王卜用言马丁由门之羽二工豕大六巴厂山曰力土耳干辛八石子已臼早车广十了白水儿虫九古皿方木月寸也四三刀川女田夕雨子甲文犬目戋己又廿竹日幺弋手巳口乙五金斤上七匕戈禾米几立一尸乃止火西弓人贝小士心王卜用言马丁由门之羽二工豕大六巴厂山曰力土耳干辛八石子已臼早车广十了白水儿虫九古皿方木月寸也四三刀川女田夕雨子甲文犬目戋己又廿竹日幺弋手巳口乙五金斤上七匕戈禾米几立一尸乃止火西弓人贝小士心王卜用言马丁由门之羽二工豕

大六巴厂山曰力土耳干辛八石子已白早车广十了白水儿虫九古皿方木月寸也四三刀川女田夕雨孑甲文犬

活动5　庖丁解牛——键外字拆分

学会了所有键面字的拆分方法后，经过刻苦练习，小文和小璐已经可以使用五笔字型输入法比较熟练地输入近百个汉字了，他俩对于学习五笔字型输入法的兴趣也越来越浓厚了。这不，他俩又迫不及待地来找师傅了。

“师傅，既然有键面字，那肯定就有键外字，你现在就教教我们键外字的拆分方法吧！”小文和小璐满怀期待地对师傅说道。

看着两个跃跃欲试的弟子，师傅却平静地说：“先别着急，我现在还不知道你们俩键面字的录入速度呢，打字基础很重要，切忌急功近利，要是没有达到30字/分钟的最低要求，我是不会教你们新知识的，你们先向我汇报一下，小文你先说吧。”

小文面露喜色地回答：“师傅，我键面字混合录入的速度是35字/分钟。”小璐紧接着有点不好意思地答道：“我的速度没有小文快，只有33字/分钟，不过我一定会努力争取追上他的。”

“嗯，你们俩的速度暂时达到了我目前的最低要求，但这并不代表你们达到了出师的水平，所以你们还要继续努力练习，争取早日达到中文录入60字/分钟的出师水平。”师傅语重心长地对两位弟子说，“只要你们每天都花一定的时间坚持练习，我相信你们俩就一定会成功，那接下来就该轮到师傅教你们新知识了，你们可要仔细听哟。”

“师傅，你就放心吧，我们一定会更加努力，争取早日出师。”小文保证道。小璐则关心要学的新知识：“师傅，你今天是教我们键外字的拆分方法吗？”

师傅肯定地点了点头，说道：“是的，中文里除了94个汉字是键面字以外，其他的都属于键外字，键外字的拆分方法和键面字完全不同，会更复杂，所以你们一定要掌握键外字的拆分原则和编码规则，在学习的过程中要多琢磨、多理解、多领悟、多练习。”

“明白了，师傅！”小文和小璐异口同声地回答。

摩拳擦掌

所有汉字都是由基本字根拼合而成的，要想拆分汉字，尤其是字根总表上没有的键外字，就必须先了解组成汉字的字根之间的相互位置关系，字根之间的位置关系也称为汉字字根间的结构关系。

汉字字根间的结构关系分为四种：单、散、连、交（见图 2—5—1）。

图 2—5—1 汉字字根间的结构关系

一、单

指汉字本身单独由一个基本字根构成，即所有的键面字。如：王、山、雨、工、小等。

二、散

指汉字由两个或两个以上基本字根构成，且字根间保持一定的距离。如：江、晶、字、树、型等。

三、连

五笔字型中汉字字根间的相连关系包含以下两种情况。

1. 汉字由一个基本字根和一个单笔画相连构成。

示例见表 2—5—1：

表 2—5—1　　字根间连的关系示例

千	丿连十	正	一连止	且	月连一
尺	尸连丶	产	立连丿	不	一连小

注意：

单笔画与基本字根之间有明显间距者不认为相连。如：个、少、么、旦、幻、旧、孔、乞、鱼等。

2. 带点结构的汉字，即汉字由点和一个基本字根构成，点和基本字根之间可连可不连。也就是说，一个基本字根和孤立点之间的关系，一律视作是基本字根相连的关系，如：太、主、术、勺、斗等。

四、交

指汉字由两个或多个字根交叉叠加而构成。

示例见表 2—5—2：

表 2—5—2　　字根间交的关系示例

申	日交丨	未	二交小	夫	二交人
果	日交木	专	二交乙	东	七交小

师傅领进门

从汉字字根间的结构关系可以看出：

单的情况即全部键面字，包括键名汉字、成字字根和单笔画汉字，它们的五笔字型编码已经有了单独规定，因此无须再进行拆分，需要拆分的是其他三种结构关系。

对于散的情况，由于组成汉字的各字根之间保持间距、相互独立，使得各字根容易辨析，因而很容易拆分。

对于连、交及连交混合的汉字，由于其字根之间相互交错，导致字根不容易辨析，因而成为键外字拆分的重点和难点，是我们在拆分中要着力解决的问题。

下面的拆分原则即针对键外字中连、交及连交混合的情况。

一、键外字的拆分原则

对于连、交及连交混合字根间结构关系的键外字，在具体拆分中要注意以下拆分原则：

取大优先，兼顾直观，能散不连，能连不交

1. 取大优先

取大优先也叫做能大不小，指在各种可能的拆法中，保证按书写顺序每次都拆出尽可能大的字根。所谓尽可能大，就是指再加一笔就不能构成已知字根。

示例：平（见图 2—5—2）、无（见图 2—5—3）。

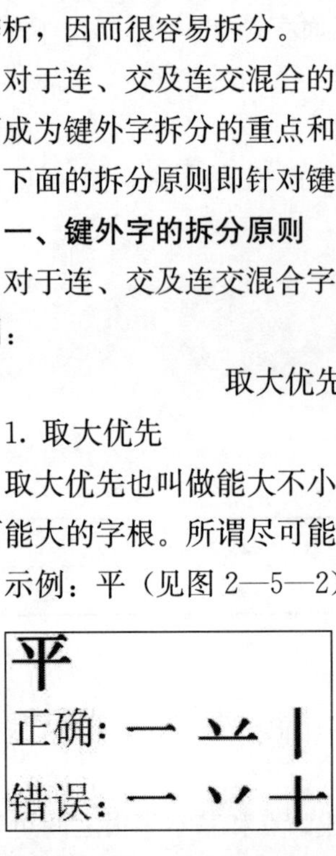

平
正确：一 丷 丨
错误：一 丷 十

图 2—5—2　例字：平

无
正确：二 儿
错误：一 一 儿、一 ナ 乚

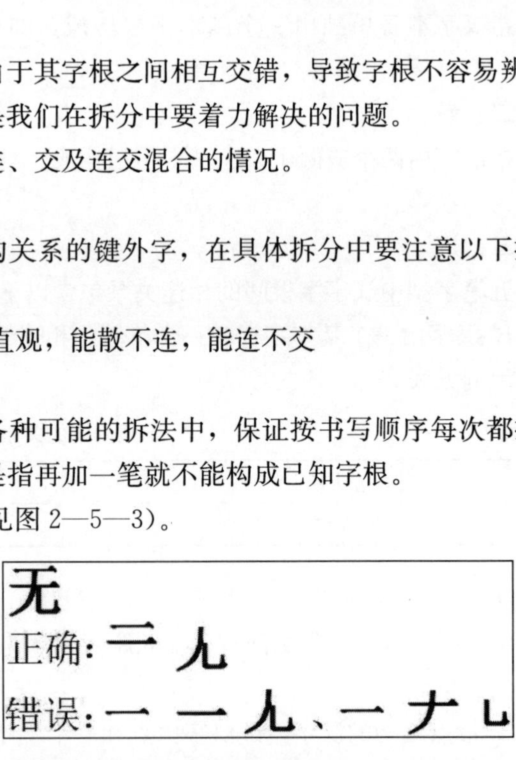

图 2—5—3 例字：无

2. 兼顾直观

指在取大优先的同时，要兼顾到拆分的直观性，从而便于联想记忆，给输入带来方便，甚至有时需要牺牲书写顺序。

示例：生（见图 2—5—4）、因（见图 2—5—5）。

生
正确：丿 龶
错误：𠂉 丨 一

图 2—5—4 例字：生

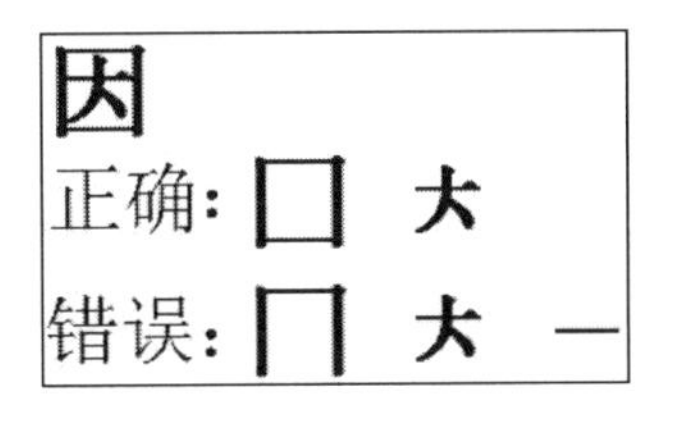

图 2—5—5 例字：因

3. 能散不连

能散不连包含以下两个含义：

（1）当两个字根相连时，只要字根不是单笔画，则一律视为散的关系。

示例：占，占可以拆分为“卜”和“口”两个字根，从直观上看，这两个字根应该是连的关系，但由于“卜”和“口”都不是单笔画，因此它们是散的关系。

（2）如果一个字的多个字根能按散的关系拆分，就不要按连的关系拆分。即一个字最好能拆分成不含单笔画的多个字根，因为根据含义，只有单笔画与基本字根之间的关系才视为连，没有了单笔画，这个字的字根之间就是散的关系了。

示例：午（见图 2—5—6）。

午
正确：𠂉 十
错误：丿 干

图 2—5—6 例字：午

4. 能连不交

指如果一个字的多个字根能按连的关系拆分，就不要按交的关系拆分，因为一般来说，连比交更为直观。

示例：天（见图 2—5—7）、开（见图 2—5—8）。

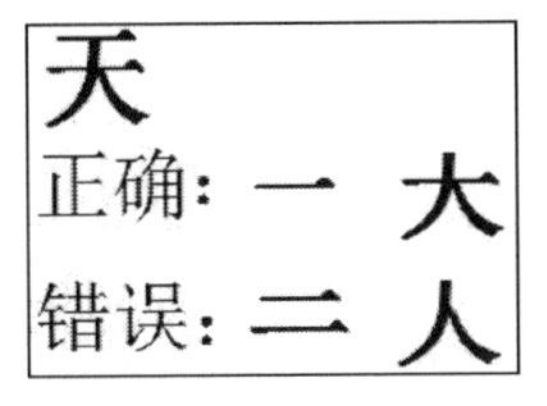

图 2—5—7 例字：天

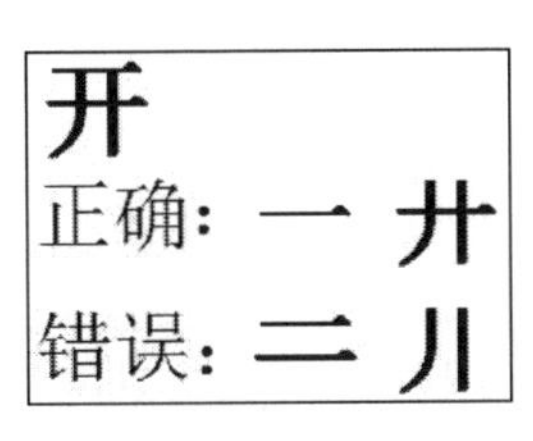

图 2—5—8 例字：开

散、连和交的关系非常重要，因为字根之间的关系，决定了汉字的字形，而字形在本项目活动 6 中将会用到，因此在键外字的拆分中一定要正确理解字根之间的关系，正确拆分键外字。

二、键外字的编码规则

键外字从其字根构成数量来看，可以分为二根字、三根字、四根字和多根字，因此键外字的编码规则自然包含二根字、三根字、四根字和多根字的编码规则。

只要在遵循键外字拆分原则的前提下，再按照下面介绍的编码规则进行编码，就能正确地输入键外字（注：不包含需要添加末笔识别码的键外字，末笔识别码见活动 6）。

1. 二根字编码规则

二根字＝第 1 个字根码＋第 2 个字根码＋空格

示例：夫（见图 2—5—9）、引（见图 2—5—10）。

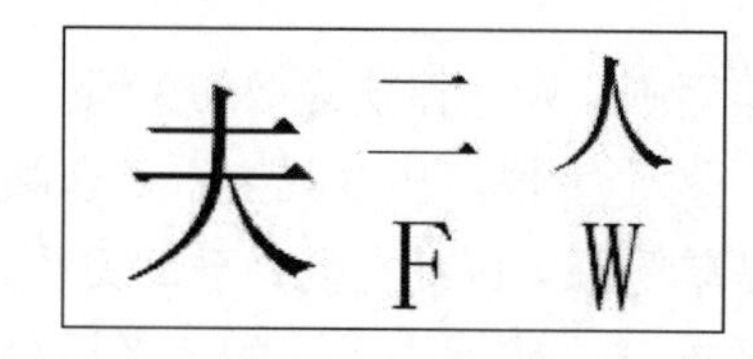

图 2—5—9　例字：夫

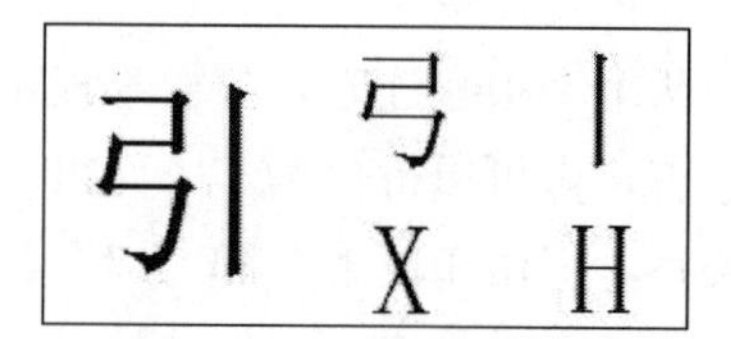

图 2—5—10　例字：引

2. 三根字编码规则

三根字＝第 1 个字根码＋第 2 个字根码＋第 3 个字根码＋空格

示例：辰（见图 2—5—11）、论（见图 2—5—12）。

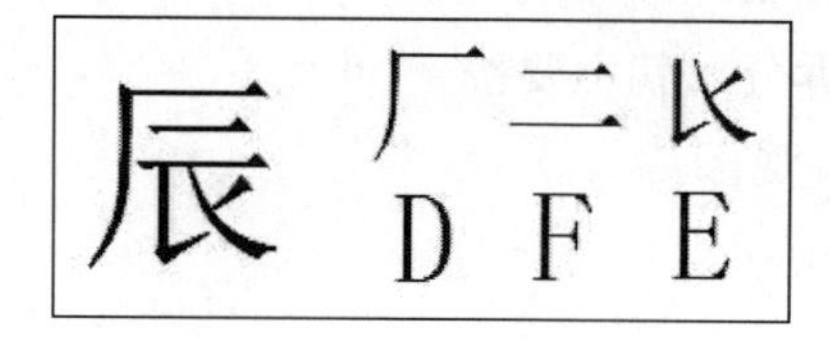

图 2—5—11　例字：辰

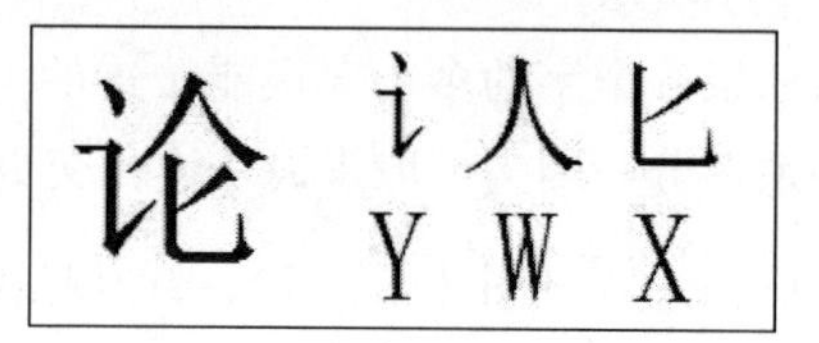

图 2—5—12　例字：论

3. 四根字编码规则

四根字＝第 1 个字根码＋第 2 个字根码＋第 3 个字根码＋第 4 个字根码

示例：氛（见图 2—5—13）、都（见图 2—5—14）。

图 2—5—13 例字：氛

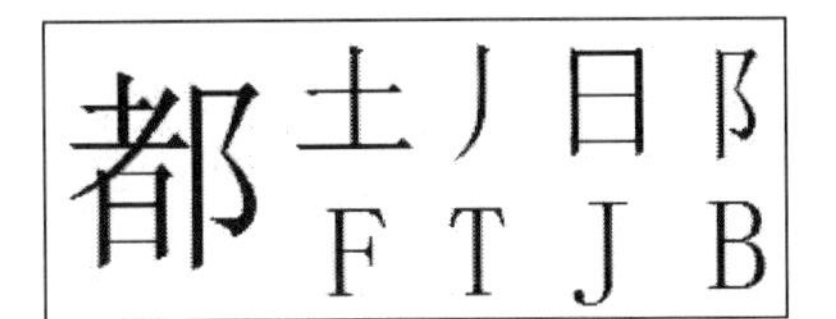

图 2—5—14 例字：都

4. 多根字编码规则

多根字＝第 1 个字根码＋第 2 个字根码＋第 3 个字根码＋末字根码

示例：编（见图 2—5—15）、孩（见图 2—5—16）。

图 2—5—15 例字：编

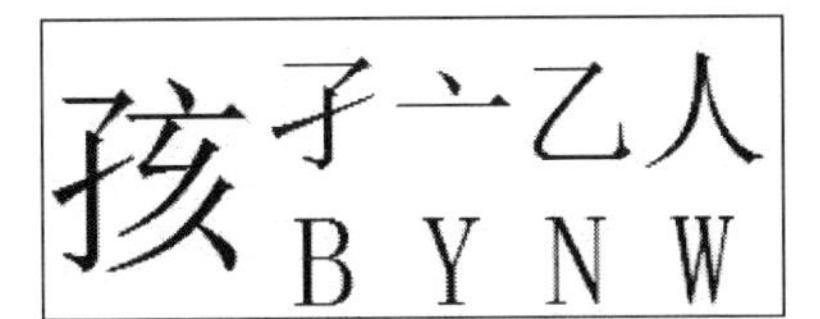

图 2—5—16 例字：孩

修行靠个人

要求：在上机练习前先写出下列汉字的编码，然后通过上机练习加以验证，之后再通过反复的上机练习（10～20 遍）达到巩固和熟练掌握的目的。

一、二根字练习

习（　　　　）生（　　　　）业（　　　　）历（　　　　）

史（　　　　）化（　　　　）他（　　　　）们（　　　　）

军（　　　　）东（　　　　）北（　　　　）下（　　　　）

左（　　　　）右（　　　　）内（　　　　）外（　　　　）

衣（　　　　）针（　　　　）线（　　　　）果（　　　　）

具（　　　　）本（　　　　）灯（　　　　）皮（　　　　）

肝（　　　　）胃（　　　　）骨（　　　　）肋（　　　　）

肛（　　　　）细（　　　　）天（　　　　）多（　　　　）

少（　　　　）出（　　　　）入（　　　　）支（　　　　）

久（　　　　）呆（　　　　）灵（　　　　）关（　　　　）

开（　　　　）阴（　　　　）阳（　　　　）放（　　　　）

取（　　　　）折（　　　　）及（　　　　）夫（　　　　）

孙（　　　　）男（　　　　）姑（　　　　）妈（　　　　）
奶（　　　　）伯（　　　　）色（　　　　）红（　　　　）
赤（　　　　）朱（　　　　）安（　　　　）包（　　　　）
边（　　　　）代（　　　　）邓（　　　　）丰（　　　　）
冯（　　　　）耿（　　　　）归（　　　　）胡（　　　　）
吉（　　　　）季（　　　　）纪（　　　　）江（　　　　）
节（　　　　）乐（　　　　）李（　　　　）林（　　　　）
刘（　　　　）娄（　　　　）卢（　　　　）吕（　　　　）
罗（　　　　）宁（　　　　）钱（　　　　）区（　　　　）
权（　　　　）全（　　　　）闰（　　　　）时（　　　　）
汪（　　　　）卫（　　　　）肖（　　　　）信（　　　　）
叶（　　　　）于（　　　　）籽（　　　　）因（　　　　）
仍（　　　　）并（　　　　）且（　　　　）时（　　　　）
间（　　　　）世（　　　　）秋（　　　　）旬（　　　　）
分（　　　　）冰（　　　　）困（　　　　）吧（　　　　）
听（　　　　）寻（　　　　）找（　　　　）打（　　　　）
辽（　　　　）宁（　　　　）汉（　　　　）明（　　　　）
昌（　　　　）太（　　　　）台（　　　　）半（　　　　）

二、三根字练习

动（　　　　）至（　　　　）阶（　　　　）种（　　　　）
而（　　　　）着（　　　　）起（　　　　）政（　　　　）
战（　　　　）性（　　　　）体（　　　　）合（　　　　）
图（　　　　）新（　　　　）论（　　　　）些（　　　　）
还（　　　　）形（　　　　）想（　　　　）点（　　　　）
育（　　　　）其（　　　　）件（　　　　）压（　　　　）
组（　　　　）数（　　　　）那（　　　　）治（　　　　）
系（　　　　）者（　　　　）意（　　　　）次（　　　　）
但（　　　　）接（　　　　）情（　　　　）运（　　　　）
质（　　　　）指（　　　　）活（　　　　）众（　　　　）
很（　　　　）根（　　　　）团（　　　　）别（　　　　）
总（　　　　）任（　　　　）更（　　　　）将（　　　　）
识（　　　　）先（　　　　）老（　　　　）复（　　　　）

完（ ）劳（ ）轮（ ）做（ ）
集（ ）号（ ）温（ ）即（ ）
研（ ）坚（ ）据（ ）织（ ）
花（ ）受（ ）求（ ）况（ ）
精（ ）界（ ）品（ ）层（ ）
清（ ）确（ ）究（ ）状（ ）
再（ ）际（ ）试（ ）布（ ）
议（ ）除（ ）齿（ ）济（ ）
效（ ）配（ ）话（ ）差（ ）
敌（ ）始（ ）施（ ）华（ ）
均（ ）存（ ）板（ ）许（ ）
非（ ）亚（ ）适（ ）讲（ ）
黄（ ）易（ ）削（ ）严（ ）
径（ ）英（ ）快（ ）坏（ ）
武（ ）助（ ）却（ ）首（ ）
府（ ）周（ ）贫（ ）朝（ ）
医（ ）呢（ ）稻（ ）范（ ）

三、四根字练习

都（ ）致（ ）律（ ）船（ ）
探（ ）零（ ）够（ ）辉（ ）
挖（ ）登（ ）津（ ）播（ ）
耐（ ）型（ ）追（ ）掌（ ）
望（ ）甚（ ）岭（ ）膜（ ）
察（ ）传（ ）勤（ ）被（ ）
貌（ ）毒（ ）磨（ ）救（ ）
岛（ ）命（ ）影（ ）造（ ）
桑（ ）资（ ）翻（ ）留（ ）
筑（ ）两（ ）善（ ）脚（ ）
贵（ ）念（ ）壁（ ）游（ ）
拿（ ）啥（ ）围（ ）照（ ）
域（ ）速（ ）建（ ）摩（ ）
洞（ ）谬（ ）荷（ ）剪（ ）

斜（ ）炼（ ）冷（ ）镇（ ）
燃（ ）觉（ ）含（ ）势（ ）
使（ ）怎（ ）制（ ）期（ ）
筒（ ）愿（ ）道（ ）臂（ ）
您（ ）铜（ ）脉（ ）抓（ ）
热（ ）爬（ ）教（ ）剩（ ）

四、多根字练习

喊（ ）感（ ）常（ ）题（ ）
鼓（ ）满（ ）该（ ）霉（ ）
愈（ ）赞（ ）废（ ）核（ ）
穗（ ）塔（ ）州（ ）锤（ ）
遵（ ）敏（ ）整（ ）赢（ ）
射（ ）骗（ ）编（ ）孩（ ）
遗（ ）盛（ ）塞（ ）疑（ ）
歌（ ）键（ ）篇（ ）酿（ ）
繁（ ）饲（ ）额（ ）裂（ ）
猪（ ）溶（ ）槽（ ）糟（ ）
褐（ ）割（ ）擦（ ）缝（ ）
穿（ ）雅（ ）警（ ）譬（ ）
露（ ）领（ ）版（ ）献（ ）
遭（ ）偏（ ）耗（ ）塑（ ）
辅（ ）龄（ ）寨（ ）韩（ ）
熙（ ）凳（ ）徽（ ）腐（ ）

灵丹妙药

一、见字就拆

键外字是最多的一类汉字，也是我们日常生活中用得最多的一类汉字。为了能尽快掌握键外字的拆分原则和编码规则，在学习初期要养成见字就拆的好习惯，比如看到报纸、书籍、杂志上的字，或者走在路上看到路牌、广告牌上的字，就有意识地做口头拆分练习。只有会拆的字越多，上机操作的时候才会越快。

二、随身准备一个小记录本

在学习键外字拆分的初级阶段，难免会遇到一些自己不会拆分的汉字，这时，可以随身准备一个小记录本，将自己不会拆分的汉字随时记录在本子上，回去以后将正确的编码写在相应汉字的后面并弄明白拆分道理，这样随着时间的推移，不会拆分的汉字就会越来越少了。

三、勤练巧练

在进行键外字拆分练习时，对于初学者而言，一定要注意做到勤练巧练，而不是死练瞎练。具体可以参考以下几点建议：

- 重视上机操作前的“书面练习”和“强记”这两个环节。“书面练习”的目的在于深化理论知识，并在上机操作前弄清楚一个字为什么要这样拆分；“强记”的目的在于对一时还不知道如何拆分的字先记住其编码，方便上机时录入，之后再逐渐理解其拆分方法。
- 对于同一个练习内容要反复多次上机练习，千万不能只练一遍两遍就放过去。要记住，单调的重复对于提高技艺是最好的训练。这里的反复多次至少也不能少于 10 次，最好能达到 20 次，只有这样，才能通过训练形成条件反射，从而做到快速录入。
- 到目前为止，大家对于一些常用汉字基本上可以进行五笔录入了，此时，不妨在日常需要进行文字录入的时候有意识地使用五笔字型输入法，比如在编辑文档、上网聊天、发邮件、写微博时等。虽然开始速度会比较慢，甚至还要和其他中文输入法进行切换使用，但是只要坚持下去，对于提高自己的拆字能力，以及五笔字型录入的速度，都是有着极大好处的。

过关斩将

要求：反复练习以下每个部分的测试内容，直到达到该部分的目标后，才能进入下一部分的测试，全部测试完成后可以进入活动 6。

第一关：二根字、三根字混合录入

目标：在 8 分钟内完成，即达到 30 字/分钟。

习范生稻业呢历医史朝化贫他周们府军首东却北助下武左坏右快内英外径衣严针削线易果黄具讲本适灯亚皮非肝许胃板骨存肋均肛华细施天始多敌少差出话入配支效久济呆齿灵除关议开布阴试阳际放再取状折究及夫确孙清男层姑品妈界奶精伯况色求红受赤花朱织安据包坚边研代即邓温丰号冯集耿做归轮胡劳吉完季复纪老江先节识乐将李更林任刘总娄别卢团吕根罗很宁众钱活指区质权运全情闰接时但汪次卫意肖者信系叶治于那籽数因组仍压并件且其时育间点世想秋形旬还分些冰论困新吧图听合寻体找性打战辽政宁起汉着明而昌种太阶台至半动

第二关：四根字、多根字混合录入

目标：在10分钟内完成，即达到30字/分钟。

都腐致徽律凳船熙探韩零寨够龄辉辅挖塑登耗津偏播遭耐献型版追领掌露望譬甚警岭雅膜穿察缝传擦勤割被褐貌糟毒槽磨溶猪救裂岛额命饲影繁造酿桑篇资键翻歌留疑筑塞两盛善遗脚孩贵编念骗壁射游赢拿整啥敏围遵照锤域州速塔建穗摩核洞废谬赞荷愈剪霉斜该炼满冷鼓镇题燃常觉感含喊势剩使教怎爬制热期抓筒脉愿铜道您臂都腐致徽律凳船熙探韩零寨够龄辉辅挖塑登耗津偏播遭耐献型版追领掌露望譬甚警岭雅膜穿察缝传擦勤割被褐貌糟毒槽磨溶猪救裂岛额命饲影繁造酿桑篇资键翻歌留疑筑塞两盛善遗脚孩贵编念骗壁射游赢拿整啥敏围遵照锤域州速塔建穗摩核洞废谬赞荷愈剪霉斜该炼满冷鼓镇题燃常觉感含喊势剩使教怎爬制热期抓筒脉愿铜道您臂

第三关：键面字、键外字混合录入

目标：在16分钟内完成，即达到30字/分钟。

范生都因照目习稻腐组锤戋业呢致仍域己历医徽压州又史朝律并速廿化贫凳件塔竹他周船且建日们府熙其穗幺军首探时摩弋东却韩育核手北助零间洞巳下武寨点废口左坏够世谬乙右快龄想赞五内英辉秋荷金外径辅形愈斤衣严挖旬剪上针削塑还霉七线易登分斜匕果黄耗些该戈具讲津冰炼禾本适偏论满米灯亚播困冷几皮非遭新吧立肝许耐鼓图一胃板献镇听尸骨存型题合乃肋均版燃寻止肛华追常体火细施领觉找西天始掌感性弓多敌露含打人少差望喊战贝出话譬势辽小入配甚剩政士支效警使宁心久济岭教起王呆齿雅怎汉卜灵除膜用爬着关议穿制明言开布察热而马阴试缝期昌丁阳际传抓种由放再擦筒太门取状勤脉阶之折究割愿台羽及夫被铜至二确孙褐道半工清男貌您动豕层姑糟臂大品妈毒六界奶槽巴精伯磨厂况色溶山求红猪曰受赤救力花朱裂土织安岛耳据包额千坚边命辛研代饲八即邓影石温丰繁子号冯造已集耿酿白做归桑早轮胡篇车劳吉资广完季键十复纪翻了老江歌白先节留水识乐疑儿将李筑虫更林塞九任刘两古总娄盛皿别卢善方团吕遗木根罗脚月很宁孩寸众钱贵也活指编四区质念三权运骗刀全情壁川闰接射女时但游田汪次赢夕卫意拿雨肖者整孑信系啥甲叶治敏文于那围犬籽数遵

活动6　一锤定音——末笔识别码

这天，小文和小璐又来到师傅的书房，向师傅汇报近期练习的进展情况。

“师傅，按照你上次给我们的‘灵丹妙药’去做，我觉得我的进步确实很大，已经顺利地‘过关斩将’了。”小文兴奋地向师傅汇报着。小璐也点头道：“是啊，我也是这样去做的，也感觉收获很大，我现在跟朋友网上聊天都在尝试用五笔字型输入法了，尽管现在速度比朋友慢，但我相信只要我努力，总有一天会超过他们的。”

看着两个热情不减的弟子，师傅欣慰地频频点头，说道："嗯，很好很好，其实你们还有一味灵丹妙药可以使用。"

"是什么呀？快告诉我们吧！"两人迫不及待地问道。

"俗话说棋逢对手，你们俩的速度差不多，正是这样的对手，那么你们可以每天都抽出一点时间进行一下比赛，打打擂台，这种持续的良性竞争对于你们提高录入水平是大有裨益的，这也是每个学文字录入的人都可以借鉴的方法之一。"师傅徐徐说道。

"嗯，有道理，"小璐看着小文说道，"每次我们都是到师傅这里来时才比拼一下，从明天开始，我们每天都比试一下怎么样？""没问题，就这么定了。"小文干脆地应道。

"唉，对了，师傅，"小璐似乎想起了什么，转身对师傅说道："我在网上聊天时，发现有些汉字我明明是按照正确的拆分方法拆分的，可就是打不出来，比如说'备''问''去'等，还有好些字也是这样，这是什么原因呢？"

"嗯，我也发现了这个问题，正摸不着头脑呢。"小文也附和道。

"问得好，"师傅笑呵呵地说："这也正是我今天要教给你们的一个新知识——末笔识别码。"

摩拳擦掌

在我们接触的众多汉字中，虽然组成每个字的笔画各不相同，但从字形结构上却可以将它们分为三类，即左右型、上下型、杂合型（或混合型），它们的字形及对应的代码和例字见表 2—6—1。

表 2—6—1　　汉字的三种字形

代码	字形	例字
1	左右	科 限 杨 情 指 结 封 准 湘 树
2	上下	星 青 雷 字 药 想 华 意 莫 算
3	杂合	电 出 包 里 头 丑 凶 这 团 乘

一、左右型汉字

左右型汉字包括以下两种情况。

- 双合字：指两个部分分列左右，中间有明显的界线。如科、限、杨等。
- 三合字：指三个部分分列左中右，如准、湘、树等；或一个部分单独占据左右的一边，另外两个部分合起来占据另一边，如指、结、封等。

二、上下型汉字

上下型汉字包括以下两种情况。

● 双合字：指两个部分分列上下，中间有明显的界线。如星、青、雷等。

● 三合字：指三个部分分列上中下，如意、莫、算等；或一个部分单独占据上下的一边，另外两个部分合起来占据另一边，如药、想、华等。

三、杂合型汉字

指组合成整字的各部分之间没有简单明确的上下左右型关系，如头、里、包、这、团等。

当我们在计算机上输入汉字时，对于有些汉字，除了要键入组成该汉字的字根外，还要告诉机器这些键入的字根是以什么方式排列的，即要补充字形信息。

师傅领进门

一、末笔识别码

大家都知道，拼音输入法的缺点是重码很多，所以打不快，不适合盲打。那么五笔字型输入法有没有重码现象呢？答案当然是肯定的。五笔字型输入法产生重码的原因主要有以下两个：

● 组成汉字的字根相同，但字形结构不同。如吧、邑，它们的字根编码都为“KC”。

● 组成汉字的字根不同，但字根处在同一个键位上。如：历、奋，它们的字根编码都为“DL”。

尽管五笔字型输入法的重码率非常低，仅有万分之三，但是为了更好地避免重码，提高录入效率，五笔字型输入法引入了末笔交叉识别码，简称末笔识别码，即对于不足 4 个字根且会产生重码的键外字，在依次敲击字根后，最后补击一个末笔识别码。

末笔识别码的代码为两位数字，分别代表其所在的区号（十位）和位号（个位），区号和位号共同组成末笔识别码所在键位的区位码，具体组成如下：

末笔识别码＝末笔画代码（十位）＋字形代码（个位）

其中，

● 末笔画代码（区号）为：横 1、竖 2、撇 3、捺 4、折 5。

● 字形代码（位号）为：左右型 1、上下型 2、杂合型 3。

根据组成末笔识别码的末笔画代码和字形代码的区位组合关系，可以得到可能的全部末笔识别码共有 15 个，分别为每个区前三位的字母，如图 2—6—1 所示。

末笔画为撇3 杂合型3 33 E	末笔画为撇3 上下型2 32 R	末笔画为撇3 左右型1 31 T	末笔画为捺4 左右型1 41 Y	末笔画为捺4 上下型2 42 U	末笔画为捺4 杂合型3 43 I
末笔画为横1 杂合型3 13 D	末笔画为横1 上下型2 12 F	末笔画为横1 左右型1 11 G	末笔画为竖2 左右型1 21 H	末笔画为竖2 上下型2 22 J	末笔画为竖2 杂合型3 23 K
C	末笔画为折5 杂合型3 53 V	末笔画为折5 上下型2 52 B	末笔画为折5 左右型1 51 N	M	,

图 2—6—1　末笔识别码

为了使取码简单，关于末笔画有如下规定：

- 末字根为“力、刀、九、七、匕”等时，其末笔画为折。
- 带“囗”的全包围字（如固、圆）、带“辶、廴”的半包围字（如连、廷），以及带“戈”的半包围字（如戒、戎），约定以被包围部分的末笔作为整个字的末笔画。
- “我、戋、成”等字的末笔画为撇。

关于字形有如下规定：

- 全包围及半包围型汉字，其字形为杂合型，如固、连、廷、戒、尿、肩、疗、庐、厌、虏、句等。
- 字根间属连和交结构关系的汉字，其字形为杂合型，如头、自、井等。
- 字根间属散结构关系的汉字，其字形可以为左右型或上下型，如枉、卡等。

示例见表 2—6—2。

表 2—6—2　　　　**末笔识别码示例**

例字	末笔画	末笔画代码	字形	字形代码	末笔识别码
扛	一	1	左右型	1	11（G）
草	丨	2	上下型	2	22（J）
栈	丿	3	左右型	1	31（T）
厌	丶	4	杂合型	3	43（I）
旯	乙	5	上下型	2	52（B）

二、末笔识别码编码规则

根据五笔字型输入法中汉字编码最多不超过 4 码的规定，可以推断出只有二根字和三根字才有可能需要添加末笔识别码，因此得到它们的编码规则如下。

1. 二根字末笔识别码编码规则

二根字＝第 1 个字根码＋第 2 个字根码＋末笔识别码＋空格

示例见表 2—6—3。

表 2—6—3　　二根字末笔识别码编码示例

例字	第 1 个字根	第 2 个字根	末笔画，字形	编码
扛	扌	工	一，左右型	RAG
章	立	早	丨，上下型	UJJ
厌	厂	犬	丶，杂合型	DDI

2. 三根字编码规则

三根字＝第 1 个字根码＋第 2 个字根码＋第 3 个字根码＋末笔识别码

示例见表 2—6—4。

表 2—6—4　　三根字末笔识别码编码示例

例字	第 1 个字根	第 2 个字根	第 3 个字根	末笔画，字形	编码
捂	扌	五	口	一，左右型	RGKG
票	西	二	小	丶，上下型	SFIU
廷	丿	土	廴	一，杂合型	TFPD

需要说明的是，所有二根字和三根字都可以通过添加末笔识别码进行编码和输入，但事实上，很多需要添加末笔识别码的字是简码字，因而我们也就无须考虑对这些字添加末笔识别码的问题了。

在中国国家标准简体中文字符集（GB 2312—1980）收录的 6 763 个汉字中，只有近 440 个汉字真正需要添加末笔识别码，本书的附录二部分列出了这些汉字的编码，供大家对照查阅。

修行靠个人

一、书面练习

要求：在上机练习前先完成下面的书面练习，加强印象，然后再上机练习，这样练习效果会更好一些。

1. 写出下列汉字的末笔画代码

汉字	末笔画代码	汉字	末笔画代码	汉字	末笔画代码	汉字	末笔画代码
昏		酥		弄		伏	
孜		勾		驰		盏	
尔		弗		未		凹	
午		农		捏		坊	
冈		页		悼		尤	
击		妒		市		闲	
秧		元		茸		尺	

2. 写出下列汉字的字形代码

汉字	字形代码	汉字	字形代码	汉字	字形代码	汉字	字形代码
井		酉		她		仅	
杀		庙		刁		迫	
泪		芯		庐		杆	
状		固		肚		奸	
闷		蚂		云		灭	
锌		冒		枚		泉	
什		位		鱼		庄	

3. 写出下列汉字的末笔识别码（用字母表示）

汉字	末笔识别码	汉字	末笔识别码	汉字	末笔识别码	汉字	末笔识别码
栗		汹		丹		矿	
酥		闯		腮		泉	
飞		沂		辜		舌	
秃		忙		尤		垃	
勺		闸		抗		卉	
痈		匆		浅		美	
钡		企		赶		触	
惊		应		正		拌	
童		丘		章		吾	
轧		苗		粒		亦	

二、末笔识别码上机练习

要求：在练习初期可以适当打开键盘帮助，以便在练习中进一步理解末笔识别码的编码规则，在熟悉之后则要关闭键盘帮助。下面的上机练习内容需反复练习 10～20 遍，直到熟练掌握为止。

艾皑岸凹叭扒笆疤把坝泵柏败拌剥卑钡狈叉备铂仓草厕叉岔忏扯彻尘程驰尺斥愁仇丑臭床闯辞触囱歹待丹悼等笛狄翟刁叮冬斗抖杜肚妒兑讹厄尔洱铒伐乏钒坊肪仿访飞吠奋忿粪封孚拂伏弗付父讣改甘杆竿赶秆冈皋杲告恭汞勾苟辜咕沽蛊故固刮挂圭旱汗夯豪亨弘户幻皇惶回茴卉昏荤霍击讥伎芰剂荠忌佳贾钾笺肩奸茧贱见涧饯溅疖秸劫戒诫巾竞今筋仅京惊井酒巨句眷卷抉诀钧君卡刊看尻抗亢苦库框矿旷亏垃兰泪厘莅笠栗溧疠粒利隶连凉晾掠疗吝漏芦庐虏仑玛吗码蚂麦忙卯冒枚眉美闷孟苗庙灭茗闽牡亩沐芳柰尿捏涅牛农弄疟呕判刨匹票迫粕扑朴钋栖奇企气乞泣讫千扦仟浅巧羌茄妾怯青琼丘酋蛆去泉雀冉壬仁刃戎茸冗汝杀晒腮杉钐汕扇尚勺舌申声升圣什矢屎仕市谁私宋诵酥粟岁她坍叹讨套贴汀廷童头秃徒吐推驮洼万丸亡枉旺妄唯未位蚊纹紊问沃吾捂毋午忤迕伍勿悟昔硒矽汐虾匣闲香湘乡翔享屑泄芯忻囟锌杏刑兄汹朽玄穴血驯丫岩阎厌唁彦秧羊佯仰杳舀耶曳沂艺邑亦异羿翌音尹页应拥佣痈蛹酉尤疣昱元圆云芸孕誉闸札扎盏章丈仗瘴兆召皂砧正置痔钟仲舟诌肘住爪庄壮状坠谆啄卓孜仔自汁走足

灵丹妙药

一、巧用【Z】键

【Z】键在五笔字型输入法中叫做“学习键”或“万能键”，一般在对某字根有疑问时使用。当无法确定某字的末笔识别码时，如果查五笔字型编码手册会比较麻烦，此时可以用【Z】键代替末笔识别码输入，然后再从提示行中点击所需要的字，同时对照该字后面的代码和最后一个末笔识别码，仔细体会一下取这个代码的道理。只要这样做，用不了多久，就能基本掌握末笔识别码的规则。

二、词组帮忙

这个方法适用于学会了词组输入的方法之后。当不确定某个字的末笔识别码时，可以输入一个含有该字的词组，例如：不确定“诫”字的末笔识别码，就输入“告诫”，然后去掉“告”字即可，这样不仅输入了“诫”字，还避开了这个字的末笔识别码。当然，本方法仅限于临时帮忙，事后还是要及时掌握这个字的编码，以免影响录入速度。

过关斩将

要求：反复练习以下每个部分的测试内容，直到达到该部分的目标以后，才能进入下一部分的测试，全部测试完成后可以进入活动 7。

第一关：左右型末笔识别码录入

目标：在 6 分钟内完成，即达到 30 字/分钟。

皑叭扒把坝柏忏败拌剥钡狈铂扯彻程驰仇辞触待悼狄叮抖杜肚妒讹洱铒伐钒坊肪仿访吠封拂伏付讣改杆秆咕沽故刮挂汗弘幻惶讥伎剂佳钾奸贱涧饯溅秸劫仅惊酒抉诀钧刊抗框矿旷垃泪溧粒利凉晾掠漏玛吗码蚂忙卯枚牡沐捏涅呕判刨粕扑朴钋栖泣讫扦仟浅巧怯琼蛆仁汝晒腮杉钐汕什仕谁私诵酥她坍叹讨贴汀徒吐推驮洼枉旺唯位蚊纹沃捂忤伍悟硒矽汐虾湘翔泄忻锌刑汹朽驯唁秧佯仰耶沂拥佣蛹札扎仗砧钟仲诌肘住状谆啄孜仔汁

第二关：上下型末笔识别码录入

目标：在 5 分钟内完成，即达到 30 字/分钟。

艾岸笆泵卑备仓草岔尘愁臭等笛翟冬兑尔奋忿粪孚父竿皋杲告恭汞苟辜蛊圭旱夯豪亨皇茴卉昏荤霍芰荠忌贯笺茧见竞今筋京眷卷卡看亢苦兰莅笠栗吝芦仑麦冒美孟苗茗亩芳柰弄票奇企气乞羌茄妾青去泉雀茸冗杀尚声圣矢市宋粟岁套童秃妥紊吾午昔香享屑芯杏兄玄穴岩彦羊杳舀艺邑亦异羿翌音页昱元云芸孕誉盏章召皂置坠卓走足

第三关：杂合型末笔识别码录入

目标：在 4 分钟内完成，即达到 30 字/分钟。

凹疤叉厕叉尺斥丑床闯囱歹丹刁斗厄乏飞弗甘赶冈勾固户回击肩疖戒诫巾井巨句君尻库亏厘疠隶连疗庐虏眉闷庙灭闽尿牛农疟匹迫千丘酋冉壬刃戎扇勺舌申升屎廷头万丸亡未问毋迕勿匣闲乡囟血丫阎厌曳尹应痈酉尤疣圆闸丈瘴兆正痔舟爪庄自

第四关：末笔识别码全录入

目标：在 15 分钟内完成，即达到 30 字/分钟。

艾皑岸凹叭扒笆疤把坝泵柏败拌剥卑钡狈叉备铂仓草厕叉岔忏扯彻尘程驰尺斥愁仇丑臭床闯辞触囱歹待丹悼等笛狄翟刁叮冬斗抖杜肚妒兑讹厄尔洱铒伐乏钒坊肪仿访飞吠奋忿粪封孚拂伏弗付父讣改甘杆竿赶秆冈皋杲告恭汞勾苟辜咕沽蛊故固刮挂圭旱汗夯豪亨弘户幻皇惶回茴卉昏荤霍击讥伎芰剂荠忌佳贯钾笺肩奸茧贱见涧饯溅疖秸劫戒诫巾竞今筋仅京惊井酒巨句眷卷抉诀钧君卡刊看尻抗亢苦库框矿旷亏垃兰泪厘莅笠栗溧疠粒利隶连凉晾掠疗吝漏芦庐虏仑玛吗码蚂麦忙卯冒枚眉美闷孟苗庙灭茗闽牡亩沐芳柰尿捏涅牛农弄疟呕判刨匹票迫粕扑朴钋栖奇企气乞泣讫千扦仟浅巧羌茄妾怯青

琼丘酋蛆去泉雀冉壬仁刃戎茸冗汝杀晒腮杉钐汕扇尚勺舌申声升圣什矢屎仕市谁私宋诵酥粟岁她坍叹讨套贴汀廷童头秃徒吐推驮洼万丸亡枉旺妄唯未位蚊纹紊问沃吾捂毋午忤迕伍勿悟昔硒矽汐虾匣闲香湘乡翔享屑泄芯忻囟锌杏刑兄汹朽玄穴血驯丫岩阎厌唁彦秧羊佯仰杳舀耶曳沂艺邑亦异羿翌音尹页应拥佣痈蛹酉尤疣昱元圆云芸孕誉闸札扎盏章丈仗瘴兆召皂砧正置痔钟仲舟诌肘住爪庄壮状坠谆啄卓孜仔自汁走足

活动 7　精益求精——拆字技巧

小文和小璐现在对五笔字型输入法的学习热情很高，字根已经练习得很熟练了，基本的输入规则师傅也已经教了。这天他们一到师傅的书房，就急切地问道："师傅，从现在开始，我们可以练习文章输入了吧。"

师傅看着他们俩，笑眯眯地说："我先考考你们，出几个字给你们拆拆看，怎么样？"

"没问题。"小文和小璐异口同声地回答。

"鸟、出、敝、身、凹、凸这几个字怎么拆"师傅问。

小文和小璐盯着这几个字，想了好长时间，一齐摇着头，盯着师傅："这几个字有点怪，字根很难确定，我们不知道怎么拆，师傅教教我们吧。"

师傅对他俩说："我们今天就是要学习这些字的拆法，我们把这些字叫难拆字。难拆字是五笔字型输入法学习过程中的一只拦路虎，它最大的特点就是'怪'，我们今天的任务就是从"怪"中找出规律，让这只拦路虎乖乖地给我们让路。"

摩拳擦掌

所谓难拆字，是指大家在拆分汉字时，对一部分汉字的字根难以确定，不知道该怎样拆好；有些汉字的书写规则不太明确，也容易造成难拆字。

师傅领进门

我们学习五笔字型输入法的过程都是边练边学，经常会遇到不会输入的"难拆字"。遇到这种情况，主要是因为笔顺不对或拆分方法有问题。我们将难拆字分为如下几种类型：

一、字根不易识别型

有很多汉字结构连为一体，对于初学者而言，很难将这类汉字拆分为正确的字根进行输入，需要在学习的过程中进行积累，掌握并记忆这些汉字的拆分方法。

示例：凹（见图 2—7—1）、凸（见图 2—7—2）、鸟（见图 2—7—3）、甘（见图 2—7—4）、韦（见图 2—7—5）、曹（见图 2—7—6）。

图 2—7—1 例字：凹

图 2—7—2 例字：凸

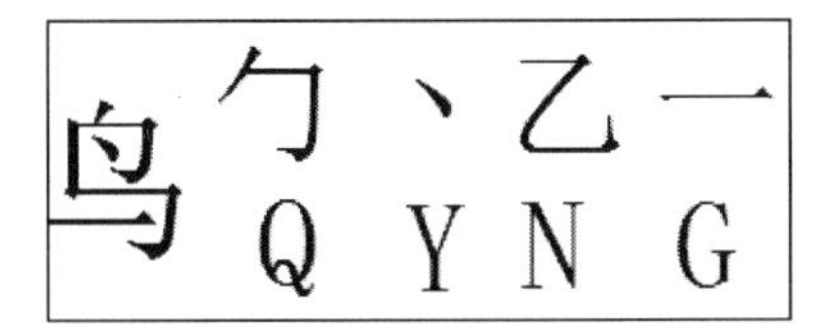

图 2—7—3 例字：鸟

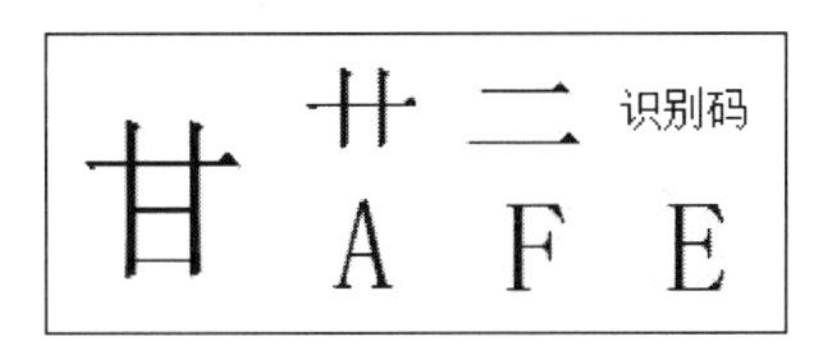

图 2—7—4 例字：甘

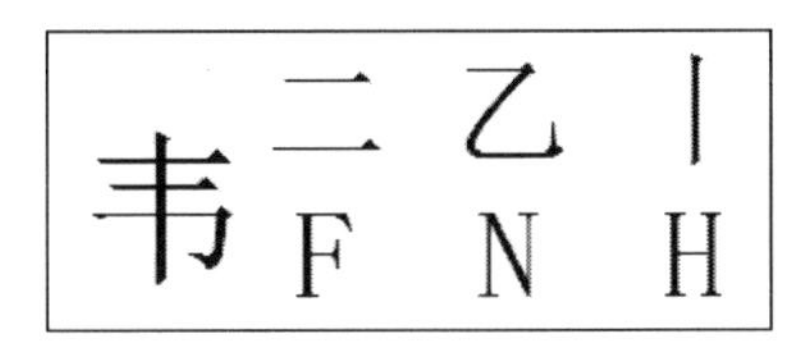

图 2—7—5 例字：韦

图 2—7—6 例字：曹

这些汉字的结构特点是每个汉字中都有一些字根是交或连在一起，初学者很难将它们正确拆分。上述类型的汉字有很多，大家一定要善于积累。

二、简单笔画型

有些汉字是由不同笔画组成，在拆字过程中，将一、二三、末四个笔画拆分出来即可。

示例：州（见图 2—7—7）、片（见图 2—7—8）

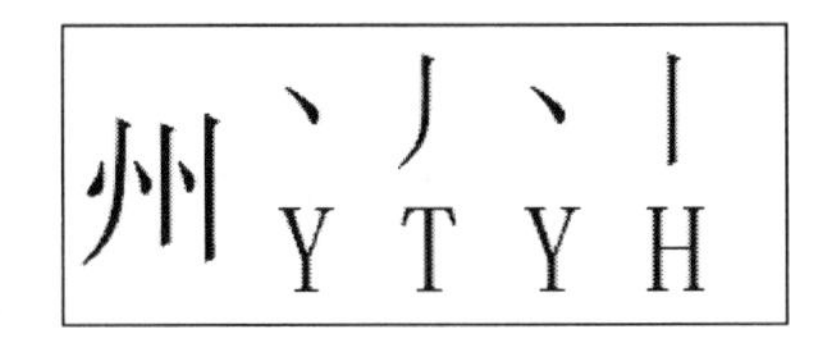

图 2—7—7 例字：州

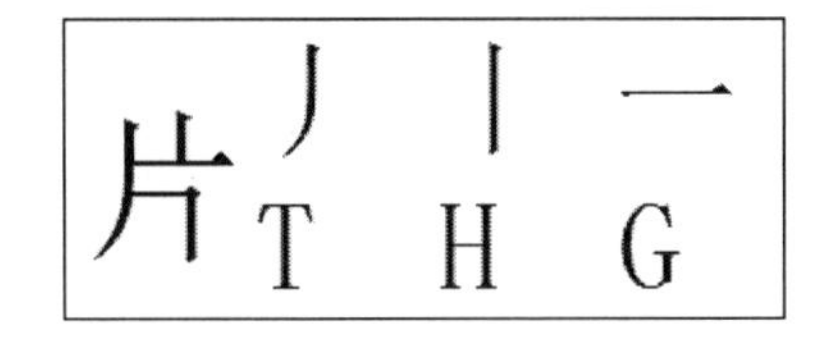

图 2—7—8 例字：片

三、易错笔画型

不少人会因为错误的书写，把一个汉字拆分为错误的字根而导致五笔字型输入法无法正确输入。

示例：沛（见图2—7—9）、害（见图2—7—10）、黄（见图2—7—11）、整（见图2—7—12）

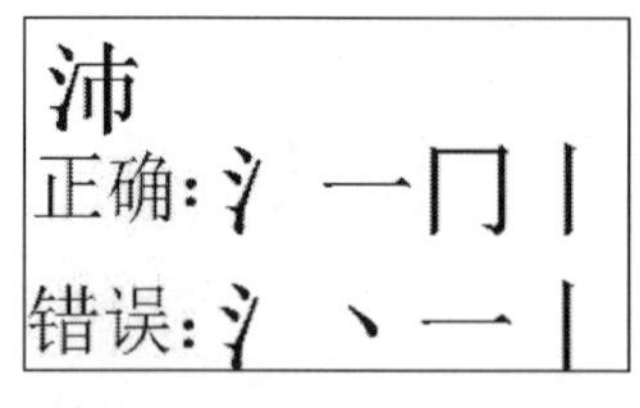

图2—7—9 例字：沛

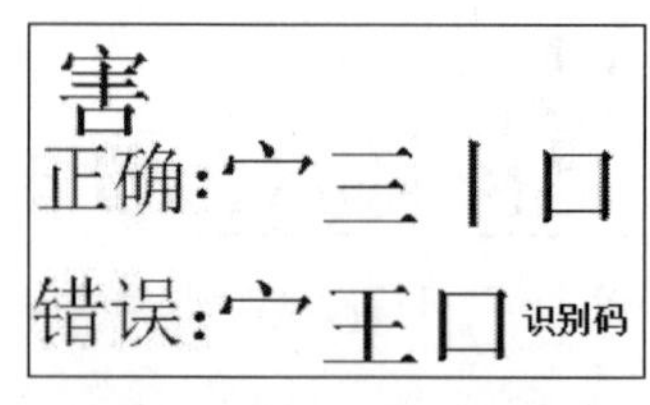

图2—7—10 例字：害

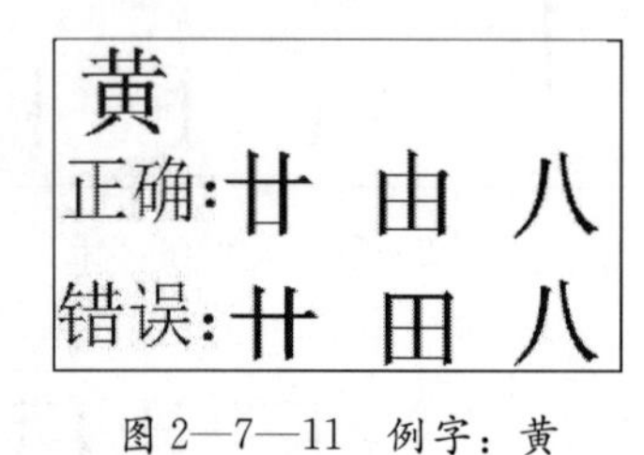

图2—7—11 例字：黄

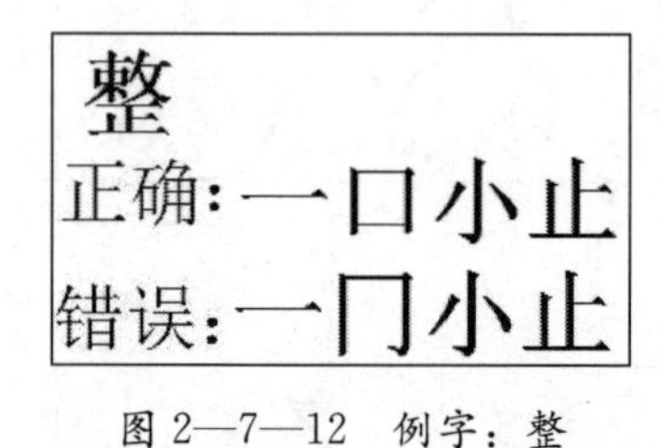

图2—7—12 例字：整

四、字根易混淆型

有些汉字无法输入是由于不同字的互相混淆引起的，我们应将这些汉字有意识地区别并记录下来。

例如“未”（FII）和“末”（GS），“天”（GD）和“夫”（FW），“已”（NNNN）和“己”（NNGN）及“巳”（NNGN）等

修行靠个人

要求：反复练习以下汉字的输入，注意下列汉字中有些汉字的字根相异性，并加以归纳。

一、书面练习

要求：在上机练习前先写出下列汉字的编码，加强印象，然后再上机练习，这样练习效果会更好一些。

凹（ ）凸（ ）丧（ ）丑（ ）釜（ ）

其（ ）期（ ）基（ ）斯（ ）甚（ ）

勘（ ）甘（ ）泔（ ）某（ ）谋（ ）

块（ ）快（ ）决（ ）牙（ ）呀（ ）

雅（　　　　）穿（　　　　）既（　　　　）尧（　　　　）烧（　　　　）
挠（　　　　）浇（　　　　）晓（　　　　）翘（　　　　）予（　　　　）
预（　　　　）序（　　　　）矛（　　　　）茅（　　　　）柔（　　　　）
亥（　　　　）该（　　　　）孩（　　　　）核（　　　　）那（　　　　）
哪（　　　　）娜（　　　　）殷（　　　　）所（　　　　）瓜（　　　　）
爪（　　　　）反（　　　　）段（　　　　）锻（　　　　）追（　　　　）
阜（　　　　）寒（　　　　）寨（　　　　）塞（　　　　）赛（　　　　）
井（　　　　）进（　　　　）非（　　　　）互（　　　　）缘（　　　　）
彝（　　　　）贯（　　　　）母（　　　　）每（　　　　）敏（　　　　）
梅（　　　　）海（　　　　）曹（　　　　）槽（　　　　）遭（　　　　）
糟（　　　　）豹（　　　　）豺（　　　　）貌（　　　　）俄（　　　　）
哦（　　　　）饿（　　　　）鹅（　　　　）物（　　　　）特（　　　　）
牡（　　　　）韦（　　　　）伟（　　　　）玮（　　　　）缶（　　　　）
缸（　　　　）缺（　　　　）卸（　　　　）制（　　　　）舞（　　　　）
兆（　　　　）逃（　　　　）桃（　　　　）脊（　　　　）东（　　　　）
陈（　　　　）练（　　　　）炼（　　　　）低（　　　　）底（　　　　）
诋（　　　　）抵（　　　　）砥（　　　　）甫（　　　　）哺（　　　　）
浦（　　　　）铺（　　　　）脯（　　　　）傅（　　　　）膊（　　　　）
溥（　　　　）簿（　　　　）切（　　　　）彻（　　　　）砌（　　　　）
倾（　　　　）革（　　　　）鞋（　　　　）靶（　　　　）鞑（　　　　）
勒（　　　　）绊（　　　　）套（　　　　）肆（　　　　）髻（　　　　）
髯（　　　　）鬃（　　　　）舟（　　　　）般（　　　　）船（　　　　）
盘（　　　　）丹（　　　　）钱（　　　　）线（　　　　）贱（　　　　）

二、上机练习

要求：在上机练习前先写出下列汉字的编码，然后通过上机练习加以验证，之后再通过反复的上机练习（10～20遍）达到巩固和熟练掌握难拆字的目的。

祭（　　　　）蔡（　　　　）察（　　　　）癸（　　　　）葵（　　　　）
免（　　　　）晚（　　　　）兔（　　　　）逸（　　　　）鬼（　　　　）
魏（　　　　）魅（　　　　）魔（　　　　）乌（　　　　）鸟（　　　　）
鸡（　　　　）岛（　　　　）曷（　　　　）喝（　　　　）渴（　　　　）
葛（　　　　）拜（　　　　）看（　　　　）差（　　　　）着（　　　　）
善（　　　　）卑（　　　　）碑（　　　　）稗（　　　　）睥（　　　　）

婢（　　　）业（　　　）亦（　　　）迹（　　　）变（　　　）
孪（　　　）兼（　　　）谦（　　　）廉（　　　）亚（　　　）
严（　　　）赤（　　　）郝（　　　）赫（　　　）余（　　　）
途（　　　）涂（　　　）叙（　　　）气（　　　）氧（　　　）
氛（　　　）敝（　　　）弊（　　　）憋（　　　）鳖（　　　）
瞥（　　　）撇（　　　）蔽（　　　）刺（　　　）枣（　　　）
策（　　　）束（　　　）速（　　　）柬（　　　）耕（　　　）
秉（　　　）重（　　　）垂（　　　）熏（　　　）隶（　　　）
逮（　　　）肃（　　　）庚（　　　）史（　　　）吏（　　　）
使（　　　）更（　　　）乘（　　　）剩（　　　）出（　　　）
击（　　　）陆（　　　）崇（　　　）即（　　　）既（　　　）
很（　　　）银（　　　）良（　　　）身（　　　）射（　　　）
躬（　　　）躺（　　　）躯（　　　）躲（　　　）承（　　　）
函（　　　）亟（　　　）丞（　　　）式（　　　）或（　　　）
武（　　　）载（　　　）栽（　　　）戴（　　　）哉（　　　）
成（　　　）戊（　　　）咸（　　　）戚（　　　）臧（　　　）
片（　　　）州（　　　）收（　　　）沛（　　　）害（　　　）
黄（　　　）勇（　　　）整（　　　）天（　　　）夫（　　　）
午（　　　）牛（　　　）矢（　　　）失（　　　）未（　　　）
末（　　　）疗（　　　）连（　　　）迫（　　　）登（　　　）

灵丹妙药

一、建议大家在拆分难拆字时，得到正确答案后把它记录下来，并分类整理出自己专用的疑难字表，仔细分析这些字是如何拆分的，自己原先拆分错误的主要原因是什么，并将相同类型的难拆字归为一类。

二、注意本身为成字字根的汉字和一般汉字输入规则的差别，并熟练掌握成字字根的输入方法。

成字字根的输入规则：报户口＋首笔笔画＋次笔笔画＋末笔笔画。

例如斤（RTTH），甲（LHNH），犬（DGTY），川（KTHH）等。

三、疑难字之间相同字根进行归纳。我们在进行汉字输入时，发现很多疑难字之间有部分相同的字根，我们应将这些汉字进行整理。

例如，其（ADW）期（ADWE）基（ADWF）斯（ADWR）甚（ADWN）勘

(ADWL)，它们的相同字根为“艹”“三”“八”；敝（UMIT）弊（UMIA）憋（UMIN）鳖（UMIG）瞥（UMIH），它们相同的字根为“丷”“冂”“小”；射（TMDF）躬（TMDX）躺（TMDK）躯（TMDQ）躲（TMDS），它们相同的字根为“丿”“冂”“三”等。

这种类型的汉字有很多，且字根都不易识别，我们在平时的练习过程中要不断地积累，以达到事半功倍的效果。

过关斩将

要求：反复练习以下每部分的测试内容，必须在 8 分钟之内完成，即达到 30 字/分钟，才能进入下一部分的测试。

第一关：常用难拆字录入 1

拜凹翱靶耙霸傲稗版拌伴半绊豹碑悲卑
辈敝弊鞭彪鳖憋斌濒秉拨博搏膊哺埠簿
蚕槽策豺搀谗厂乘承虫丑锄楚处舜斯撕
嘶肆似巳捶锤垂寸撮歹耽丹岛第碘典叼
懂毒犊锻蛾峨鹅娥饿耳贰筏伐乏阀飞夫
敷釜脯赋阜父缚感皋革根跟耕更羹躬辜
瓜罐惯贯瑰鬼裹喊撼憾毫黑很哼亨喉猴
弧互缓换昏豁惑击棘脊既荚颊甲兼缄拣
减践贱见溅建降浇嚼睫戒斤谨韭九厩救
臼舅疽倦眷卷撅攫抉爵君俊卡慨慷糠靠
克亏喇来澜谰揽览懒缆滥榔狼廊朗浪牢
勒镰敛脸炼练粮撩聊僚疗潦镣临龄榴六
窿鹿旅率卵麦忙矛茂貌美寐妹门冕免勉
娩缅面蔑灭鸣末某拇姆哪乃囊蔫年鸟孽

第二关：常用难拆字录入 2

凝扭纽脓浓疟哦鸥藕偶爬排牌徘湃派判
叛耪砰抨片撇瞥频苹萍瓶婆破莆圃普妻
气遣歉羌墙蔷且禽球求曲躯龋犬缺炔榷
雀群瓤壤攘嚷扰绕壬刃绒伞丧陕膳善烧
舌身升盛剩失尸士世市手戌耍甩睡瞬搜
艘塑溯肃穗祟糖躺掏逃套藤腾誊舔腆挑

眺跳廷彤头凸兔退吞屯鸵唾瓦袜歪挽亡
问瓮呜钨毋五午舞侮坞戊西夕霞暇峡乡
晓楔卸行兄羞锈戌血熏讯迅鸦呀丫芽牙
蚜雅哑亚讶焉蜒演彦央鸯秧痒漾曳夜夷
遗胰疑姨彝已乙肄赢庸用由迂隅予御渊
冤缘曰凿枣早乍炸窄毡栈丈仗整正止舟
株茱铢珠属瞩爪足亍丏廿禺氏兆谏尥尬
尴弋制追幺自缶竹曹粤甲乙乘臧魔肆癸

活动 8　一马当先——录入一级简码

经过前段时间的认真学习，小文和小璐已经掌握了五笔字型输入法的基本规则，并且还在努力地进行难拆字练习。他们现在正跃跃欲试，等着师傅给他们布置任务呢。

师傅说："别着急，先考考你们，这、经、要这几个字怎么输入？"

"简单，这：ypi，经：xcag，要：svf。"两人争先恐后地回答。

师傅说："很好，你们遵循的是五笔字型基本输入规则，这样输入肯定没问题，但为了使五笔的输入更快，我们给出现频率最高的 25 个汉字定义为一级简码，其中'这、经、要'就属于一级简码。"

摩拳擦掌

在五笔字型方案中，将人们在说话和写文章时，使用频率最高的 25 个汉字精选出来，放在 A—Y 25 个键位上，根据键位上的字根形态特征，每键安排一个极为常用的高频汉字，这类字只要输入一个字母键加一个空格键就可以了，这类字就称为"一级简码"（见图 2—8—1）。

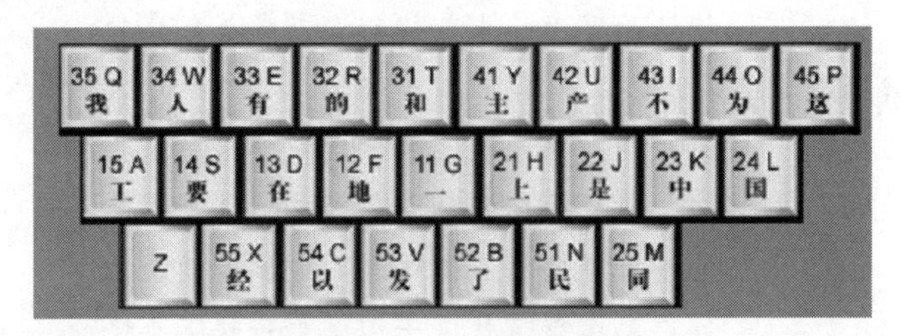

图 2—8—1　一级简码键盘图

师傅领进门

一级简码（又称高频字）有 25 个，是用一个字母键和一个空格键作为一个汉字的编码。

以下是从一区到五区的一级简码及所对应的字母键。

1 区：一（G）　地（F）　在（D）　要（S）　工（A）

2 区：上（H）　是（J）　中（K）　国（L）　同（M）

3 区：和（T）　的（R）　有（E）　人（W）　我（Q）

4 区：主（Y）　产（U）　不（I）　为（O）　这（P）

5 区：民（N）　了（B）　发（V）　以（C）　经（X）

助记短句：我是一有为的中国人，要和以上不同地民工在这经主产发了。

修行靠个人

要求：反复进行以下每个部分的内容输入练习，直到每个区的一级简码背熟以后，才能进入下一个部分的输入练习。

一、从 1 区至 5 区按顺序进行训练

要求：反复练习以下每个一级简码，输入的同时将一级简码与相应键位对应，达到巩固和掌握每个一级简码的目的。

一地在要工上是中国同和的有人我主产不为这民了发以经一地在要工上是中国同和的有人我主产不为这民了发以经一地在要工上是中国同和的有人我主产不为这民了发以经一地在要工上是中国同和的有人我主产不为这民了发以经一地在要工上是中国同和的有人我主产不为这民了发以经一地在要工上是中国同和的有人我主产不为这民了发以经一地在要工上是中国同和的有人我主产不为这民了发以经一地在要工上是中国同和的有人我主产不为这民了发以经

二、1 区至 5 区的一级简码混合训练

要求：反复练习以下每个一级简码，输入的同时将一级简码与相应键位对应，达到巩固和掌握每个一级简码的目的。

地有不为这人有不以同了有一上中在国要一同发了一上不有国是国工同有不为人我这有不同民了发以经为这有我不有中在为工不为有同民的为人这我上一以不有为一要这有的不一为有这在这一不人为工这有的不一上在不为在同发民发以不这我有上一在国有不和主不人这发同主一有不要这在同民了发以有为一上在为人上一在不为经；

工在上我不有这不上以民了有为上不有为一同以有不为经上一有不工要在中国有不为人一上不有为要这在一是国有不要

灵丹妙药

注意：虽然一级简码也可以通过多编码方式录入，但会大大影响录入速度，因此大家必须牢记在心，从而快速而准确地录入（见表2—8—1）。

表2—8—1　　一级简码汉字的一级编码与多级编码对照表

一级简码汉字	一级编码	多级编码
一	G	G
地	F	FBN
在	D	DHFD
要	S	SVF
工	A	AAAA
上	H	HHG
是	J	JGH
中	K	KHK
国	L	LGY
同	M	MGK
和	T	TKG
的	R	RQY
有	E	DEF
人	W	WWW
我	Q	TRN
主	Y	YG
产	U	YTE
不	I	GI
为	O	YLYI
这	P	YPI
民	N	NA
了	B	BNH
发	V	NTC
以	C	NYW
经	X	XCA

过关斩将

要求：反复练习以下每个部分的测试内容，必须在指定时间内完成，即达到 80 字/分钟后，才能进入下一个部分的测试。

第一关：分区练习（3 分钟）

一地在要工上是中国同和的有人我主产不为这民了发以经一地在要工上是中国同和的有人我主产不为这民了发以经一地在要工上是中国同和的有人我主产不为这民了发以经一地在要工上是中国同和的有人我主产不为这民了发以经一地在要工上是中国同和的有人我主产不为这民了发以经一地在要工上是中国同和的有人我主产不为这民了发以经一地在要工上是中国同和的有人我主产不为这民了发以经一地在要工上是中国同和的有人我主产不为这民了发以经一地在要工上是中国同和的有人我主产不为这民了发以经一地在要工上是中国同和的有人我主产不为这民了发

第二关：混合区练习（5 分钟）

国同和的有人我主产不为这民了发以经不发以有为这我民了发不为在是发以有人为我这上一不为有和主的人为同以不发了以不为有人为不一上在中不有和主产不为发以不为要为同以不为人为同民不为有人为在中国同和的有人我主产不为这民了一不为有和主同以不发了以不这我民了发不为在为在中国同和的有人我主产不为这民要为同以不为人为同以不为人为同民不为有人为在中国同和的有人我主产不为这这民了发以经不发以有为这我民了发不为在是发以有人为我这上一不为有和主的人为同以不发了以不为有人为不一上在中不有了发不为在为在中国同和的有人我主产不为这民要为同以不为人为同以不为人为同民不为有人为在中国同和的有人我主产不为这这民了发以经不发以有为这我民了发不为在是发以有人为我这上一不为有和主的人为同以不发了以一地在要工上是中国同和的有人我主产不为这民了发以经不发以有为这我民了发不为在是发以有人为我这上一不为有和主的人为同以不经一地在要工上是中

活动 9　合二为一——录入二级简码

小文和小璐已经将 25 个一级简码练习得滚瓜烂熟了，师傅笑眯眯地对他们说：“除了 25 个一级简码，还有 606 个二级简码呢。”

小文和小璐嘴巴张得老大，面面相觑：“师傅，606 个二级简码哪背得了呀，这些汉字我们直接按五笔字型的输入规则输入不就得了吗?”

师傅不紧不慢地说：“按照五笔的输入规则固然可以，但速度就慢了很多，掌握二级简码没什么窍门，就是勤学苦练，进行专题训练，熟练了，看到这个字自然而然形成条件反射，就知道按二级简码的规则输入。你们有没有信心掌握这606个二级简码呀？”

小文和小璐不约而同地大声回答：“有！”

摩拳擦掌

在五笔输入过程中，定义了使用频率最高的25个一级简码输入方法，除此之外，还定义了606个高使用频率的二级简码的输入方法（见表2—9—1），极大地提高了五笔的输入速度。

表2—9—1　　二级简码表

	G	F	D	S	A	H	J	K	L	M	T	R	E	W	Q	Y	U	I	O	P	N	B	V	C	X
G	五	于	天	末	开	下	理	事	画	现	玫	珠	表	珍	列	玉	平	不	来	★	与	屯	妻	到	互
F	二	寺	城	霜	载	直	进	吉	协	南	才	垢	圾	夫	无	坟	增	示	赤	过	志	地	雪	支	★
D	三	夺	大	厅	左	丰	百	右	历	面	帮	原	胡	春	克	太	磁	砂	灰	达	成	顾	肆	友	龙
S	本	村	枯	林	械	相	查	可	楞	机	格	析	极	检	构	术	样	档	杰	棕	杨	李	要	权	楷
A	七	革	基	苛	式	牙	划	或	功	贡	攻	匠	菜	共	区	芳	燕	东	★	芝	世	节	切	芭	药
H	睛	睦	睚	盯	虎	止	旧	占	卤	贞	睡	睥	肯	具	餐	眩	瞳	步	眯	瞎	卢	╳	眼	皮	此
J	量	时	晨	果	虹	早	昌	蝇	曙	遇	昨	蝗	明	蛤	晚	景	暗	晃	显	晕	电	最	归	紧	昆
K	呈	叶	顺	呆	呀	中	虽	吕	另	员	呼	听	吸	只	史	嘛	啼	吵	噗	喧	叫	啊	哪	吧	哟
L	车	轩	因	困	轼	四	辊	加	男	轴	力	斩	胃	办	罗	罚	较	★	辚	边	思	团	轨	轻	累
M	同	财	央	朵	曲	由	则	★	崭	册	几	贩	骨	内	风	凡	赠	峭	赕	迪	岂	邮	╳	凤	嶷
T	生	行	知	条	长	处	得	各	务	向	笔	物	秀	答	称	入	科	秒	秋	管	秘	季	委	么	第
R	后	持	拓	打	找	年	提	扣	押	抽	手	折	扔	失	换	扩	拉	朱	搂	近	所	报	扫	反	批
E	且	肝	须	采	肛	胩	胆	肿	肋	肌	用	遥	朋	脸	胸	及	胶	膛	膦	爱	甩	服	妥	肥	脂
W	全	会	估	休	代	个	介	保	佃	仙	作	伯	仍	从	你	信	们	偿	伙	★	亿	他	分	公	化
Q	钱	针	然	钉	氏	外	旬	名	甸	负	儿	铁	角	欠	多	久	匀	乐	炙	锭	包	凶	争	色	★
Y	主	计	庆	订	度	让	是	训	为	高	放	诉	衣	认	义	方	说	就	变	这	记	离	良	充	率
U	闰	半	关	亲	并	站	间	部	曾	商	产	瓣	前	闪	交	六	立	冰	普	帝	决	闻	妆	冯	北
I	汪	法	尖	洒	江	小	浊	澡	渐	没	少	泊	肖	兴	光	注	洋	水	淡	学	沁	池	当	汉	涨
O	业	灶	类	灯	煤	粘	烛	炽	烟	灿	烽	煌	粗	粉	炮	米	料	炒	炎	迷	断	籽	娄	烃	糨
P	定	守	害	宁	宽	寂	审	宫	军	宙	客	宾	家	空	宛	社	实	宵	灾	之	官	字	安	★	它
N	怀	导	居	★	民	收	慢	避	惭	届	必	怕	★	愉	懈	心	习	悄	屡	忱	忆	敢	恨	怪	尼
B	卫	际	承	阿	陈	耻	阳	职	阵	出	降	孤	阴	队	隐	防	联	孙	耿	辽	也	子	限	取	陛
V	姨	寻	姑	杂	毁	叟	旭	如	舅	妯	九	★	奶	★	婚	妨	嫌	录	灵	巡	刀	好	妇	妈	姆
C	骊	对	参	骠	戏	★	骒	台	劝	观	矣	牟	能	难	允	驻	骈	★	╳	驼	马	邓	艰	双	★
X	线	结	顷	★	红	引	旨	强	细	纲	张	绵	级	给	约	纺	弱	纱	继	综	纪	弛	绿	经	比

在二级简码表中要注意以下几点：

1. 有“╳”的为无字二码域，有3个：HB；MV；CO。

2. 有“★”的二码域可输入词组，有16个：GP不定期；FX超级大国；AO工业区；LI团党委；MK风吹草动；WP倾家荡产；QX煞费苦心；PC客观存在；NE避孕药；NS尽可能；VR忍气吞声；VW群众观点；CH能上能下；CI邓小平；CX又红又专；XS纵横驰骋。

3. 同时为一级简码字的有11个：不，地，要，中，同，主，为，这，产，民，经。

师傅领进门

二级简码字的简码和其全码的前两位相同，即只用前两个字根编码。掌握二级简码是非常重要的，因为这些字是我们日常生活中最常见的字，也就是说，汉字虽有几十万个，但常用的也只有一两千个，所以这些字的重要性就不言而喻了。

二级简码由单字的前两个字根组成，输入二级简码字时，一般只需要输入单字全码的前两个字根，后跟空格键即可。

二级简码的输入规则：第一个字根＋第二个字根＋空格键。

二级简码包括：

（1）部分不需要识别码的二根字，如庆、全、呆。

（2）部分键名汉字、成字字根、三根字、四根字、多根字，如大、五、累、辊、餐。

二级简码由于只需打前两个字根，所以可以大大加快录入速度，而且由于二级简码通常都是使用频率较高的字，所以需要熟练掌握。

修行靠个人

一、书面练习

要求：写出以下每个部分中二级简码的编码，加深印象，达到掌握和巩固二级简码的目的。

1. 5个键名汉字及20个成字字根为二级简码

（1）键名汉字

大（　　　）立（　　　）水（　　　）之（　　　）子（　　　）

（2）成字字根

二（　　　）三（　　　）四（　　　）五（　　　）六（　　　）

七（　　　）九（　　　）早（　　　）车（　　　）力（　　　）

手（　　　　）方（　　　　）小（　　　　）米（　　　　）由（　　　　）
几（　　　　）心（　　　　）也（　　　　）用（　　　　）马（　　　　）

2. 难字二级简码

涨（　　　　）艰（　　　　）限（　　　　）宽（　　　　）载（　　　　）
帮（　　　　）顾（　　　　）基（　　　　）睡（　　　　）餐（　　　　）
哪（　　　　）嘛（　　　　）啊（　　　　）笔（　　　　）秘（　　　　）
肆（　　　　）换（　　　　）曙（　　　　）最（　　　　）城（　　　　）
喧（　　　　）爱（　　　　）偿（　　　　）遥（　　　　）率（　　　　）
瓣（　　　　）澡（　　　　）煤（　　　　）降（　　　　）慢（　　　　）
避（　　　　）愉（　　　　）懈（　　　　）隐（　　　　）绵（　　　　）
楞（　　　　）弱（　　　　）贩（　　　　）晃（　　　　）宛（　　　　）
晕（　　　　）嫌（　　　　）磁（　　　　）联（　　　　）怀（　　　　）
互（　　　　）第（　　　　）毁（　　　　）菜（　　　　）赠（　　　　）
曾（　　　　）屡（　　　　）增（　　　　）瞎（　　　　）害（　　　　）
紧（　　　　）管（　　　　）得（　　　　）就（　　　　）遇（　　　　）
婚（　　　　）离（　　　　）

3. 二级简码分类

（1）学习和学科

学（　　　　）习（　　　　）学（　　　　）科（　　　　）生（　　　　）
物（　　　　）学（　　　　）业（　　　　）学（　　　　）生（　　　　）
作（　　　　）业（　　　　）历（　　　　）史（　　　　）地（　　　　）
理（　　　　）物（　　　　）理（　　　　）化（　　　　）学（　　　　）
三（　　　　）角（　　　　）

（2）称呼

称（　　　　）呼（　　　　）你（　　　　）他（　　　　）们（　　　　）
工（　　　　）商（　　　　）学（　　　　）军（　　　　）

（3）方向

方（　　　　）向（　　　　）东（　　　　）南（　　　　）北（　　　　）
中（　　　　）下（　　　　）左（　　　　）右（　　　　）前（　　　　）
后（　　　　）内（　　　　）外（　　　　）

（4）日常生活

衣（　　　　）服（　　　　）针（　　　　）线（　　　　）水（　　　　）

果（ ）大（ ）米（ ）面（ ）粉（ ）
具（ ）餐（ ）菜（ ）烟（ ）笔（ ）
本（ ）

（5）人体

皮（ ）手（ ）脸（ ）面（ ）肝（ ）
胆（ ）胃（ ）胸（ ）膛（ ）怀（ ）
牙（ ）心（ ）眼（ ）睛（ ）瞳（ ）
骨（ ）肋（ ）肛（ ）

（6）同义字和反义字

大（ ）小（ ）强（ ）弱（ ）粗（ ）
细（ ）天（ ）地（ ）多（ ）少（ ）
出（ ）入（ ）收（ ）支（ ）长（ ）
久（ ）呆（ ）灵（ ）站（ ）立（ ）
睡（ ）下（ ）关（ ）开（ ）进（ ）
出（ ）爱（ ）恨（ ）阴（ ）阳（ ）
放（ ）取（ ）给（ ）直（ ）折（ ）
曲（ ）各（ ）与（ ）及（ ）

（7）家庭

结（ ）婚（ ）离（ ）婚（ ）成（ ）
家（ ）夫（ ）妻（ ）儿（ ）子（ ）
孙（ ）子（ ）奶（ ）伯（ ）

（8）数量词

二（ ）三（ ）四（ ）五（ ）六（ ）
七（ ）九（ ）百（ ）

（9）颜色

色（ ）红（ ）赤（ ）朱（ ）绿（ ）
棕（ ）

（10）姓氏

安（ ）包（ ）边（ ）步（ ）查（ ）
车（ ）成（ ）代（ ）邓（ ）方（ ）
丰（ ）冯（ ）高（ ）耿（ ）宫（ ）
顾（ ）官（ ）管（ ）归（ ）果（ ）

胡（　　）怀（　　）吉（　　）季（　　）纪（　　）
江（　　）节（　　）经（　　）景（　　）乐（　　）
李（　　）历（　　）力（　　）林（　　）刘（　　）
龙（　　）娄（　　）卢（　　）吕（　　）罗（　　）
马（　　）米（　　）南（　　）年（　　）定（　　）
皮（　　）钱（　　）强（　　）区（　　）曲（　　）
权（　　）全（　　）闰（　　）商（　　）时（　　）
史（　　）水（　　）孙（　　）铁（　　）汪（　　）
卫（　　）闻（　　）习（　　）向（　　）肖（　　）
信（　　）燕（　　）杨（　　）阳（　　）叶（　　）
阴（　　）由（　　）于（　　）原（　　）载（　　）
曾（　　）支（　　）朱（　　）左（　　）

(11) 动植物

虎（　　）马（　　）龙（　　）驼（　　）蝗（　　）
蛤（　　）燕（　　）叶（　　）籽（　　）果（　　）

(12) 常用语

因（　　）为（　　）所（　　）以（　　）这（　　）
样（　　）式（　　）作（　　）学（　　）习（　　）
仍（　　）然（　　）并（　　）且（　　）或（　　）

(13) 时间和天气

时（　　）间（　　）早（　　）晨（　　）晚（　　）
间（　　）世（　　）纪（　　）小（　　）时（　　）
年（　　）天（　　）春（　　）秋（　　）旬（　　）
分（　　）秒（　　）冰（　　）霜（　　）

(14) 语气词

呀（　　）嘛（　　）吧（　　）啊（　　）哟（　　）

(15) 动作

听（　　）吵（　　）睡（　　）提（　　）寻（　　）
找（　　）打（　　）给（　　）啼（　　）呼（　　）

(16) 地名

安（　　）阳（　　）百（　　）色（　　）长（　　）
江（　　）辽（　　）宁（　　）大（　　）同（　　）

汉（　　　　）城（　　　　）汉（　　　　）阳（　　　　）昆（　　　　）
明（　　　　）汉（　　　　）阴（　　　　）吉（　　　　）林（　　　　）
九（　　　　）江（　　　　）九（　　　　）龙（　　　　）绵（　　　　）
阳（　　　　）南（　　　　）昌（　　　　）南（　　　　）宁（　　　　）
南（　　　　）阳（　　　　）曲（　　　　）阴（　　　　）阳（　　　　）
太（　　　　）原（　　　　）五（　　　　）台（　　　　）信（　　　　）
阳（　　　　）

二、上机练习

要求：在上机练习前先写出下列二级简码的编码，然后通过上机练习加以验证，之后再通过反复的上机练习（10～20 遍）达到巩固和熟练掌握二级简码的目的。

景暗晃显晕电最归紧昆呈叶顺呆呀中虽吕另员呼听吸只史嘛啼吵喧叫啊哪吧哟车轩因困四辊加男轴力斩胃办罗罚较边思轨轻累同财央朵曲由则崭册几贩骨内风凡赠峭迪岂邮凤生行知条长处得各务向笔物秀答称入科秒秋管秘季委么第后持拓打找年提扣押抽手折扔失换扩拉朱搂近所报扫反批且肝介保佃仙作伯仍从顾肆锭包凶争色扔失换扩拉朱搂近所报扫楷七革基苛显晕电最归紧昆呈叶承阿陈耻阳职阵不降枯林械相查瞎卢眼皮此秋管决闻妆冯北汪法类结顷红引旨强细纲张绵级给约纺弱纱继综他分公化钱针然钉氏术样档杰棕曲信

灵丹妙药

一、收集需要加补识别码的常用二根字，这样就能和不需加识别码的二根字（即二级简码）区别开，以达到掌握二根字二级简码的目的。

例如：未元杜青弄正麦歹走刊
址井圳盏吾故丈页夯厌
泵奋厄访应枉码酉栈杠
杳杏矿卡备呷呗叽杀血

二、二级简码数量较多，共有 606 个，不必死记硬背，关键是需要进行二级简码的专题训练，提高熟练度，达到见到二级简码，自然地形成条件反射，直接进行二码输入。

过关斩将

要求：以下每部分的测试内容，必须在 7 分钟内完成，即达到 40 字/分钟，才能进入下一部分的测试。

第一部分：二级简码录入 1

五于天末开下理事画现玫珠表珍列玉平不来与屯妻到互二寺城霜载直进吉协南才垢圾夫无坟增示赤过志地雪支三夺大厅左丰百右历面帮原胡春克太磁砂灰达成顾肆友龙本村枯林械相查可楞机格析极检构术样档杰棕杨李要权楷七革基苛式牙划或功贡攻匠菜共区芳燕东袄芝世节切芭药睛睦盯虎止旧占卤贞睡肯具餐眩瞳步眯瞎卢眼皮此量时晨果虹早昌蝇曙就昨蝗明蛤晚景暗晃显晕电最归紧昆呈叶顺呆呀中虽吕另员呼听吸只史嘛啼吵喧叫啊哪吧哟车轩因困四辊加男轴力斩胃办罗罚较边思轨轻累同财央朵曲由则蕲册几贩骨内风凡赠峭迪岂邮凤生行知条长处得各务向笔物秀答称入科秒秋管秘季委么第后持拓打找年提扣押抽手折扔失换扩拉朱搂近所报扫反批且肝肛胆肿肋肌用

第二部分：二级简码录入 2

遥朋脸胸及胶膛爱甩服妥肥脂全会估休代个介保佃仙作伯仍从你信们偿伙亿他分公化钱针然钉氏外旬名甸负儿铁角欠多久匀乐炙锭包凶争色主于庆订度让刘训为高放诉衣认义方说就变这记离良充率闰半关亲并站间部曾你产瓣前闪交六立冰普帝决闻妆冯北汪法尖洒江小浊澡渐没少泊肖兴光注洋水淡迷泌池当汉涨业灶类灯煤粘烛炽烟灿烽煌粗粉炮米料炒炎迷断籽娄烃定守害宁宽寂审宫军宙客宾家空宛社实宵灾之官字安它怀导居民收慢避惭届必怕愉懈心习悄屡忱忆敢恨怪尼卫际承阿陈耻阳职阵不降孤阴队隐防联孙耿辽也子限取陛姨寻姑杂毁旭如舅九奶婚妨嫌录灵巡刀好妇妈姆对参戏台劝观矣牟能难允驻驼马邓艰双线结顷红引旨强细纲张绵级给约纺弱纱继综纪弛绿经比

第三部分：二级简码录入 3

主于庆订度五于天末开下理事画现玫珠表珍列玉平不夫无坟增示赤过志地雪支三夺大厅让刘训为高放诉衣认义方说就变这记离良充率闰半估休代个介保佃仙作伯仍从顾肆锭包凶争色扔失换扩拉朱搂近所报扫楷七革基苛显晕电最归紧昆呈叶承阿陈耻阳职阵不降枯林械相查瞎卢眼皮此秋管决闻妆冯北汪法类结顷红引旨强细纲张绵级给约纺弱纱继综他分公化钱针然钉氏外旬名甸负儿铁角欠多久匀乐炙关亲并站间部曾你产瓣前闪交六立式牙顺灶呆呀中虽吕另员孤阴队隐防联孙耿或功贡攻匠菜共区芳燕东袄芝世节切芭药睛睦且肝肛胆肿协南才垢圾江小浊澡渐没少泊肖兴光注洋水淡划批用遥朋脸胸及胶膛爱甩服妥友龙本村灯煤粘烛炽烟灿烽煌粗粉炮米料炒炎迷断籽娄烃定守

第四部分：二级简码录入 4

宁之官肋量习悄忱敢恨怪尼际后持拓打年提扣押抽手折困四辊加男轴力斩办罗凤生行知条长处得各务向笔物秀答称入科秒你信们偿伙亿来与害屯妻到忆胃二灾懈寺城霜载直进吉字安吸只史嘛啼吵喧叫啊哪吧哟呼听冰普帝尖洒较朵边思审轨轻累军同财卫央曲由寂则屡宽蕲册几贩骨内风凡赠峭迪岂邮肌景暗晃肯具餐眩瞳步眯达成宙客宾

家空宛社实宵时找晨果虹迷沁池当汉涨业反肥脂全会它辽左丰百右历面帮原胡春克太磁砂灰可楞机格析极检构术样档杰棕杨李要权秘季委么第盯虎止旧占卤贞睡早昌蝇曙就昨蝗明蛤晚也子限取陛姨寻姑杂毁旭如舅九奶婚妨嫌录灵巡刀好妇妈姆对参戏台劝观矣牟能难允驻驼马邓艰双线纪弛绿经比车轩因怀导居民收慢避嘶届必怕愉心互罚宫

活动10　熟能生巧——录入常用字

小文和小璐跟着师傅学艺已经有很长时间。

一大早，他们俩就等在师傅的书房里准备开始新一天的学习了。

“师傅，我们都学习这么久了，拆字的方法也都基本学会了，怎么打字的速度就是提高不上去呢?”小文问道。

“是呀，好奇怪!”小璐附和道。

“其实，在我们日常的文章中，有一部分字使用的频率非常高，如果这些字你们练熟掌握的话，对提升你们的速度有很大帮助。”

“是哪些字?”小文和小璐马上问道。

“这些字我们通常称为常用千字，它们的使用频率在90%以上。千字中包含了键名字、一级简码、二级简码等，对它们的练习要循序渐进、练习中要注意使用简码。师傅相信练好常用字后，你们的文章录入速度一定会有大幅度提升的。”

小文和小璐对视了一眼，不约而同地大声回答：“我们一定行!”

摩拳擦掌

国家标准GB 2312—1980《信息交换用汉字编码字符集——基本集》是根据使用频率制订的。一级字库为常用字3 755个，二级字库为不常用字3 008个，一、二级字库共有汉字6 763个。一级字库中的字，使用频率达99.7%，即在现代汉语材料中的每一万个汉字中，这些字就会出现9 970次以上，其余的所有汉字也不足30次。而最常用的1 000个汉字，使用频率在90%以上，因此，这1 000字是五笔字型输入法练习中提高汉字录入速度的重中之重。

师傅领进门

千字在日常的文章中使用的频率很高，提高千字的录入速度很重要，那么怎么来练习和提速呢?

一、循序渐进

千字练习必须循序渐进。按照千字中使用频率由高到低，依次可以分为：42 字、100 字、300 字、500 字和 1 000 字。分层的常用千字练习，必须建立在自觉要求的基础上，循序渐进、步步为营、自定目标、训练到位（目标、速度、测试）。

二、注重简码

千字练习时必须注意简码的使用。因为一级简码 25 个，二级简码理论上 625 个，实际为 606 个，这两者已占用常用字的 61.2%。可见，一、二级简码占千字的比例是非常高的，练习一、二级简码的目的就是要压缩码长、提高有效键数、形成条件反射。码长是指平均打出 1 个单字的按键次数，计算公式为：码长＝总键数/总字数，总键数是指在单位时间内总的击键次数，总字数是指在单位时间内总的汉字录入数，码长的数字越大，表示文章录入过程中简码的使用情况越差。一些专用打字软件都具有统计总键数的功能，如图 2—10—1 所示。

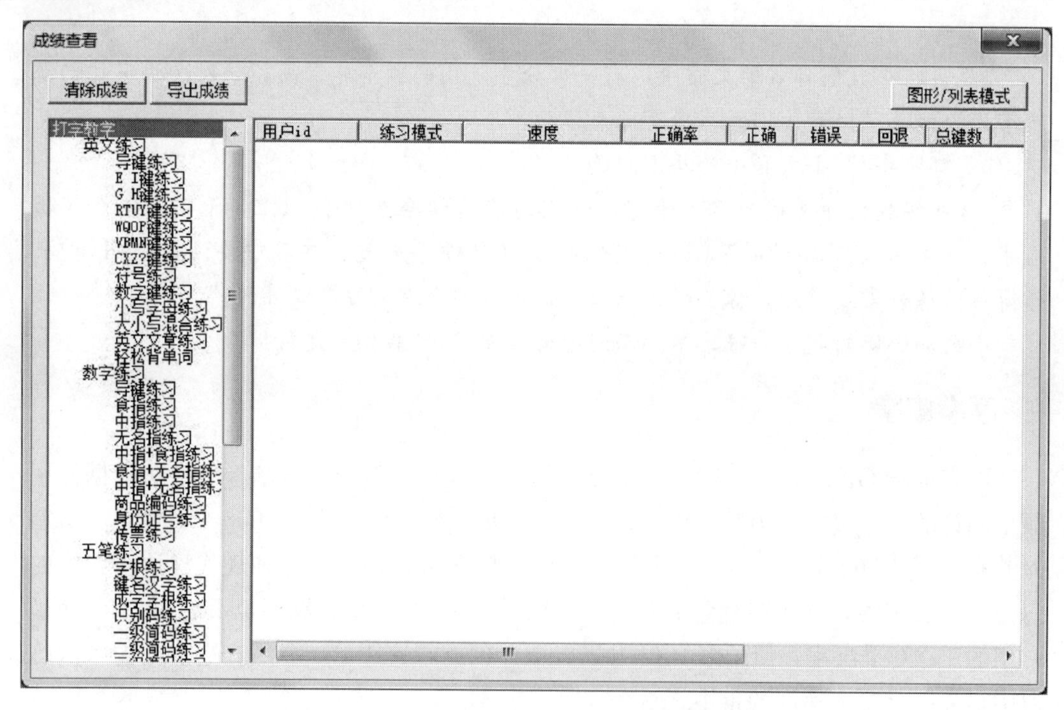

图 2—10—1　打字旋风软件中“总键数”“回退数”统计

三、科学练习

在千字的练习中要运用科学的方法，练习后要有仔细地分析，才能帮助你更快进步。

录入过程中要注意提高键速、缩小码长，降低回改率。键速是指每秒按键次数，回退率是指回退键击键次数占总击键的比例。它们之间的关系如下：

键速=总键数/时间；

回退率=（回退数/总键数）×100%。

回退数在一些专用打字软件也有统计，如图2—10—1所示。

修行靠个人

要求：反复练习以下每部分的输入内容，直到该部分的目标达成以后，才能进入下一部分的输入练习。

一、千字中42字的练习

目标：录入速度50字符/分钟，即在3分钟以内完成录入。

的一是在了不和有大这去中人上为们地个用工时要动国产以我到他会作来分生对于学下级义就年的一是在了不和有大这去中人上为们地个用工时要动国产以我到他会作来分生对于学下级义就年的一是在了不和有大这去中人上为们地个用工时要动国产以我到他会作来分生对于学下级义就年这去中人上为们地个用工时要动国产以我到他会作来分

日期	速度	键速	码长	回退率
训练体会				

二、千字中100字的练习

目标：录入速度50字符/分钟，即在2分钟以内完成录入。

的一国在人了有中是年和大业不为发会工经上地市要个产这出行作生家以成到日民来我部对进多全建他公开们场展时理新方主企资实学报制政济用同于法高长现本月定化加动合品重关机分力自外者区能设后就等体下万元社过前面

日期	速度	键速	码长	回退率
训练体会				

三、千字中 300 字的练习

目标：录入速度 30 字符/分钟，即在 10 分钟以内完成录入。

的一了是我不在人们有来他这上着个地到大里先力完间却站代员机更说就去子得也和那要下看天时过出小么起你都少直意夜比阶连车重便把好还多没为又可家学只以主会样年想能生同者干石满日决百原拿群老中十从自面前头道它后然走很像见两用她国八难早论吗根共让相研动进成回什边作对开而已些现山民候经发工向步反处记将千找争领或事命给长水几义三声于高正妈手知理眼志点心九您每风级跟笑啊孩万战二问但身方实吃做叫当住听革打呢真党全才斗马哪化太指变社似士四已所敌之最光产情路分总条白话东席次亲如究各六本思解立河爸村被花口放儿常西气五第使写军吧文运再果怎定今其书坐接应关信觉死许快明行因别飞外树物活部门无往船望新带队师结块跑谁草越字加脚

日期	速度	键速	码长	回退率
训练体会				

四、千字中 500 字的练习

1. 第一组练习

目标：录入速度 30 字符/分钟，即在 10 分钟以内完成录入。

的一了是我不在人们有来他这上着个地到大里说就去子得也和那要下看天时过出小么起你都把好还多没为又可家学只以主会样年想能生同老中十从自面前头道它后然走很像见两用她国动进成回什边作对开而已些现山民候经发工向事命给长水几义三声于高正妈手知理眼志点心战二问但身方实吃做叫当住听革打呢真党全才四已所敌之最光产情路分总条白话东席次亲如被花口放儿常西气五第使写军吧文运再果怎定许快明行因别飞外树物活部门无往船望新带队先力完间却站代员机更九您每风级跟笑啊孩万少直意夜比阶连车重便斗马哪化太指变社似士者干石满日决百原拿群究各六本思解立敌之最光产情路分总条白话东席次亲如被花口放儿风级跟笑啊孩万少直意夜比阶连车重便斗被花样

日期	速度	键速	码长	回退率
训练体会				

2. 第二组练习

目标：录入速度 30 字符/分钟，即在 10 分钟以内完成录入。

哪化太指变社似士者干石满日决百原拿群究各六本思解立河爸村八难早论吗根共让相研今其书坐接应关信觉死步反处记将千找争领或师结块跑谁草越字加脚紧爱等习阵怕月青半火法题建赶位唱海七女任件感准张团屋爷离色脸片科倒睛利世病刚且由送切星导晚表够整认响雪流未场该并底深刻平伟忙提确近亮轻讲农古黑告界拉名呀土清阳照办史改历转画造嘴此治北必服雨穿父内识验传业菜爬睡兴形量咱观苦体众通冲合破友度术饭公旁房极南枪读沙岁线野坚空收算至政城劳落钱特围弟胜教热展包歌类渐强数乡呼性音答哥际旧神座章帮啦受系令跳非何牛取入岸敢掉忽种装顶急林停息句娘区衣般报叶压母慢叔背细倒送

日期	速度	键速	码长	回退率
训练体会				

五、千字中 1 000 字综合练习

1. 第一组练习

目标：录入速度 30 字符/分钟，即在 10 分钟以内完成录入。

激找叫云互跟裂粮母练塞钢顶策双留误粒础吸阻故寸晚丝女焊攻株亲院冷彻弹错散尼盾商视艺灭版烈零室轻血倍缺厘泵察绝富城喷简否柱李望盘磁雄似困巩益洲脱投送奴侧润盖挥距触星松获独官混纪座依未突架宽冬兴章湿偏纹执矿寨责阀熟吃稳夺硬价努翻奇甲预职评读背协损棉侵灰虽矛罗厚泥辟告卵箱掌氧恩爱停曾溶营终纲孟钱待尽俄缩沙退陈讨奋械胞幼哪剥迫旋征槽殖握担仍呀载鲜吧卡粗介钻逐弱脚怕盐末阴丰编印蜂急扩伤飞域露核缘游振操央伍甚迅辉异序免纸夜乡久隶缸夹念兰映沟乙吗儒杀汽磷艰晶插埃燃欢铁补咱芽永瓦倾阵碳演威附牙斜灌欧献顺猪洋腐请透司危括脉若尾束壮暴企莱穗楚汉愈绿拖牛份染既秋遍锻玉夏疗尖井费州访吹荣铜沿替滚客召旱悟刺脑措贯藏令隙

日期	速度	键速	码长	回退率
训练体会				

2. 第二组练习

目标：录入速度30字符/分钟，即在10分钟以内完成录入。

达尔场织历花受求传口断况采精金界品判参层止边清至万确究书低术状厂需离再目海交权且儿青才证越际八试规斯近注办布门铁需走议县兵虫固除般引齿千胜细影济白格效置推空配刀叶率今选养德话查差半敌始片施响收华觉备名红续均药标记难存测士身紧液派准斤角降维板许破述技消底床田势端感往神便圆村构照容非搞亚磨族火段算适讲按值美态黄易彪服早班麦削信排台声该击素张密害候草何树肥继右属市严径螺检左页抗苏显苦英快称坏移约巴材省黑武培著河帝仅针怎植京助升王眼她抓含苗副杂普谈围食射源例致酸旧却充足短划剂宣环落首尺波承粉践府考刻靠够满夫失住枝局菌杆周护岩师举曲春元超负砂封换太模贫减阳包江扬析亩木言球朝医校古呢稻宁听唯输滑站另卫字鼓

日期	速度	键速	码长	回退率
训练体会				

3. 第三组练习

目标：录入速度30字符/分钟，即在10分钟以内完成录入。

优占促死毒圈伟季命度革而多子后自社加小机也经力线本电高量长党得实家定深法表着水理化争现所二起政三好十战无农使性前等反体合斗路图把结第里正新开论之物从当两些还天资事对批如应形想制心样干都向变关点育重其思与间内去应件日利相由压员气业代全组数果期导平各基月毛然问比或展那它最及外没看治提五解系林者米群头意只明四道马认次文通但条较克又公孔领军流入接席位情运器并习原油放立题质指建区验活众很教决特此常石强极土少已根共直团统式转别造切九你取西持总料连任志观调么七山程百报更见必真保热委手改管处巳将修支识病象先老光专几什六型具示复安带每东增则完风回南广劳轮科北打积车计给节做务被整联步类集号列温中即毫轴知研单色坚据

日期	速度	键速	码长	回退率
训练体会				

灵丹妙药

一、寻找短板，反复练习

在练习过程中有的字看到后不能马上拆，要稍作思考后，才能录入，这个字就是卡壳字；有的字练习时尽管第一时间就能拆，但连续两、三次都拆错，要反复修改。这两类字要逐一做好登记，反复练习。

二、科学练习，稳步提升

在千字的综合练习过程中要打乱字的顺序进行练习和测试。由于时间的限制，在常用千字综合练习、测试阶段，如果一直按照固定顺序就会出现前面的字非常熟，后面的字由于能打到的次数很少，依然比较陌生的情况，达不到千字录入速度整体提升的效果。

三、软件介绍

进行千字的练习可以使用“打字旋风软件”“打字高手”等软件，如图 2—10—2 是“打字旋风软件”中千字分层练习。

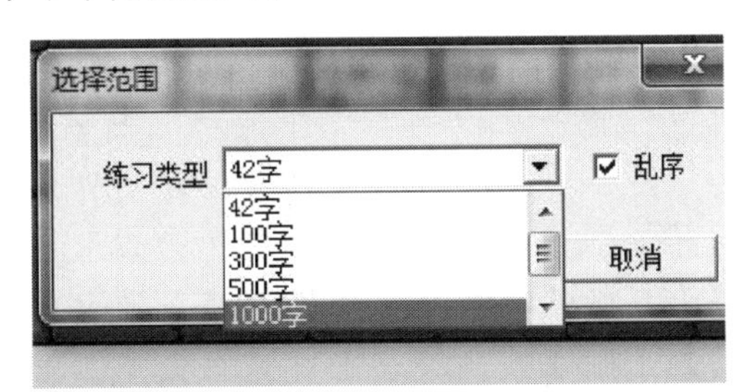

图 2—10—2　常用千字分层练习

过关斩将

要求：反复练习以下每部分的测试内容，测试时间为 12 分钟，必须达到 40 字/分钟后，才能进入下一部分的测试，全部测试完成后可以进入活动 11。

第一关：常用千字顺序录入

1. 前 500 字

的一国在人了有中是年和大业不为发会工经上地市要个产这出行作生家以成到日民来我部对进多全建他公开们场展时理新方主企资实学报制政济用同于法高长现本月定化加动合品重关机分力自外者区能设后就等体下万元社过前面农也得与说之员而务利电文事可种总该三各好金第司其从平代当天水省提商十管内小技位目起海所立已通

入量子问度北保心还科委都术使明着次将增基名向门应里美由规今题记点计去强两些表系办教正条最达特革收二期并程厂如道际及西口京华任调性导组东路活广意比投决交统党南安此领结营项情解议义山先车然价放世间因共院步物界集把持无但城相书村求治取原处府研质信四运县军件育局干队团又造形级标联专少费效据手施权江近深更认果格几看没职服台式益想数单样只被亿老受优常销志战流很接乡头给至难观指创证织论别五协变风批见究支那查张精林每转划准做需传争税构具百或才积势举必型易视快李参回引镇首推思完消值该走装众责备州供包副极整确知贸己环话反身选亚么带采王策真女谈严斯况色打德告仅它气料神率识劳境源青护列兴许户马港则节款拉直案股光较河花根布线土克再群医清速律她族历非感占续师何影功负验望财类货约艺售连纪按讯史示象养获石食抓富模始住赛客越闻央席坚

2. 后500字

份士热限米银息校均房周游千失八检足配存九命尔即防钱评复考依断范础油照段落访未额双让切须儿便空往你层低奖注黄英承远版维算破铁乐边初满病响药助致善突爱容香称购届余素请白宣健牌促培竞巴稳继紧字困刘旅声超随例担友号显却监材且春居适除红半买充陈火搞图阳六察试太什执片古七球修尽控讲排粮武预亲挥卖审措荣洲卫希店良属险曾围域令站苏龙念罗吨器汇康减习演普田班待星飞写矿轻扩言章汽靠毛终仍景置底福止离泽波兰核降训逐票菜座献钢眼损宁像苦印融独湖早予夫编换欧努著顾征升态套介送某斗状画留航派室临兵补宝略黑综云差纳密贫剧犯阿击遇岁阶烈督吃丰馆招害官树听庭另沙私针胜贷网愿托缺园假酒音巨既判输讨测读洋括筑欢刚庆久陆找楼激晚绝压故互签汉草木亩短绍迎吸警藏疗贵纷授登探索湾宏录申诉秀序顺死卡歌午孩桥喜川邓扬津温船库订练候退违否彩棉帮拿罪币角召灾妇杨奋绩虽煤免笔够永圳停奥鲜朝吴岛觉移尼急博贯拥束左细舞幅语俄奇般简拍脑债固威券追筹刻映繁伟甚饭右彻烟沿街血冲洪植誉刊玉斤救潮迅伍怎付倍顿述播励斤乎纸振旧障鼓艰呼吉男绿尚夏亏季松哈祖典韩遍夜轮板抗摄杂皮贡借幕罚伤岸扶乱曲脱践危澳童散味叶累谢孙邮雄兼微呢谁惠偿署择染答块徐鱼赞课盛延瑞怀堂

第二关：常用千字乱序录入

1. 前500字

飞外树物活部门无往船望新带队先力完间却站代员机更九您每风级跟笑啊孩万少直意夜比阶连车重便斗马哪化太指变社似士者千石满日决百原拿群究各六本思解立河爸村八难早论吗根共让相研今其书坐接应关信觉死步反处记将千找争领或师结块跑谁草越字加脚紧爱等习阵怕月青半火法题建赶位唱海七女任件感准张团屋爷离色脸片科

倒睛利世病刚且由送切星导晚表够整认响雪流未场该并底深刻平伟忙提确近亮轻讲的一了是我不在人们有来他这上着个地到大里说就去子得也和那要下看天时过出小么起你都把好还多没为又可家学只以主会样年想能生同老中十从自面前头道它后然走很像见两用她国动进成回什边作对开而已些现山民候经发工向事命给长水几义三声于高正妈手知理眼志点心战二问但身方实吃做叫当住听革打呢真党全才四已所敌之最光产情路分总条白话东席次亲如被花口放儿常西气五第使写军吧文运再果怎定收算至政城劳落钱特围弟胜教热展包歌类渐强数乡呼性音答哥际旧神座章帮啦受系令跳非何牛取入岸敢掉忽种装顶急林停息句娘区衣般报叶压母慢叔背细许快明行因别农古黑告界拉名呀土清阳照办史改历转画造嘴此治北必服雨穿父内识验传业菜爬睡兴形量咱观苦体众通冲合破友度术饭公旁房极南枪读沙岁线野坚空

2. 后500字

律雨让骨远初皮播优占促死毒圈伟季训控激找云互裂粮练塞钢策双留误粒础吸阻故寸晚丝焊攻株亲院冷彻弹错散尼盾商视艺灭版烈零室轻血倍缺厘泵察绝富喷简否柱李盘磁雄似困巩益洲脱投送奴侧润盖挥距触星松获独官混纪依未突架宽冬兴章湿偏纹执矿寨责省武培著帝仅针怎植京助升王眼她抓含苗副杂普谈围食射源例致酸充足短划剂宣环首尺波承粉践府考刻靠夫失枝局菌杆周护岩师举曲春元超负砂封换模贫减阳江扬析亩木言球朝医校稻宁唯输滑另卫鼓写刘微略范供阿某功套友限项余卷创阀熟稳侵灰虽矛罗厚泥辟卵箱掌氧恩停曾溶营终纲孟钱待尽俄缩沙退陈讨奋械胞幼剥迫旋征槽殖握担仍呀载鲜卡粗介钻逐弱怕盐末阴丰编印蜂扩伤域露核缘游振操央伍甚迅辉异序免纸夜乡久隶缸夹念兰映沟乙儒杀汽磷艰晶插埃燃欢铁补咱芽永瓦倾阵碳演威附牙斜灌欧献顺猪洋腐请透司危括脉若尾束壮暴企菜穗楚汉愈绿拖份染既秋遍锻玉夏疗尖井费州访吹荣铜沿替滚客召旱悟刺脑措贯配刀叶率选养德查差始片施响收华备红续均药标存测士身液派准斤角降维板许破述技消床田势端神圆构照容搞巴材夺硬价努翻奇甲预职评协损棉藏隙影济格效置推亚磨族段算适按值美态黄易彪班麦削信排台声击素张密害肥继右属市严径螺检左页抗苏显英称坏移约

第三关：常用千字乱序随机500字录入

战无农使性前等反体合斗路图把结第里正新开论之物从当两些还天资事对批如应形想制心样干都向变关点育重其思与间内去应件日利相由压员气业代全组数果期导平各基月毛然问比或展那它最及外没看治提五解系林者米群头意只明四道马认次文通但条较克又公孔领军流入接席位情运器并习原油放立题质指建区验活众很教决特此常石强极土少已根共直团统式转别造切九你取西持总料连任志观调么七山程百报更见必真

保热委手改管处已将修支识病象先老光专几什六型具示复安带每东增则完风回南广劳轮科北打积车计给节做务被整联步类集号列温中即毫轴知研单色坚据速防史拉世射达尔场织历花受求传口断况采精金界品判参层止边清至万确究书低术状厂需离再目海交权且儿青才证越际八试规斯近注办布门铁需走议县兵虫固除般引齿千胜细影济白格效置推空配刀叶率今选养德话查差半敌始片施响收华觉备名红续均药标记难存测士身紧液派准斤角降维板许破述技消底床田势端感往神便圆村构照容非搞亚磨族火段算适讲按值美态黄易彪服早班麦削信排台声该击素张密害候草何树肥继右属市严径螺检左页抗苏显苦英快称坏移约巴材省黑武培著河帝仅针怎植京助升王眼她抓含苗副杂普谈围食射源例致酸旧却充足短划剂宣环落首尺波承粉

活动11　事半功倍——录入词组

小文和小璐上次听师傅的话，练好了常用千字后，想去参加中英文录入擂主争霸赛，没想到师傅却说："你们先跟我比赛，如果赢了我就去比赛，如果输了继续学习。"

于是比赛开始了，小文和小璐没几个回合就败下阵来。

小文和小璐不解地问师傅："您教的方法我们都会了，您让我们练习千字我们也练了，究竟我们差在哪里呢？"

师傅笑了笑说："年轻人，不要急躁，师傅基本的方法都教过你们了，但还有一些诀窍没教完，比如今天要教你们的词组的输入方法，它也是提速的一个秘密武器。"

"所有文章都是由单字和词组组成的，五笔中的词组有二字、三字、四字和多字词。这些词有一个共同的特点，那就是四码就能打出来，当然并非所有语文中的词组在五笔中都是词组，要靠平时的积累。"师傅继续说道。

"那不是很神奇嘛！四码能出两个字、三个字、四个字、甚至多个字！"小文惊奇地说，"不过，这一定会很难吧？"

"其实词组录入一点也不难！只要有拆字的基础，掌握正确的方法，每个人都能很快掌握。"

"看来，今天我们又能掌握一些诀窍了，师傅请放心，我们一定会加油的！"小文说。

小璐说："青出于蓝而胜于蓝，我们一定努力，以超过师傅为目标！"

摩拳擦掌

词组是文章的重要组成部分，词组的快速录入是提升文章录入速度的又一诀窍，而词组的输入是五笔输入又一神奇所在，它可以四码打出两个字、也能四码打出三个字、多个字，甚至可以三码打出两个字。下面通过一段文字使用单字录入和词组录入的总码长比较，来观察词组录入的神奇之处。

示例：

中国特色社会主义理论体系，就是包括邓小平理论、“三个代表”重要思想以及科学发展观等重大战略思想在内的科学理论体系。

单字录入：字数52，总键数117，平均码长2.25。

1 1 3 2 3 3 2 2 3 3 3 3 2 1 2 3 2
中 国 特 色 社 会 主 义 理 论 体 系，就 是 包 括 邓
2 2 3 3 2 2 2 2 3 1 2 3 1 2 2 2 1
小 平 理 论、“三 个 代 表”重 要 思 想 以 及 科 学 发
3 2 4 3 2 3 3 2 3 1 2 1 2 2 2 3 3 3
展 观 等 重 大 战 略 思 想 在 内 的 科 学 理 论 体 系。

词组录入：字数52，总键数74，平均码长只有1.42。

4 4 4 4 4 4 4
中国特色 社会主义 理论体系，就是 包括 邓小平理论、“三个代表”
4 4 4 4 2 4 4 4 4 1 2 1
重要思想 以及 科学 发展 观 等 重大 战略 思想 在 内 的
4 4
科学 理论体系。

由此可以看出，上面这段文章，如果使用词组录入会比单字录入总键数减少37%，录入速度当然也就有大幅度提高了。

师傅领进门

一、二字词组的录入

二字词组的输入方法为：

首字的前两个字根＋次字的前两个字根

注意：作为词组录入时，如果词组中第一个字是一级简码，仍然要遵循以上方法，其他多字词组同样如此。

示例见表2—10—1。

表 2—10—1 二字词组

例词	首字 第一个字根	首字 第二个字根	次字 第一个字根	次字 第二个字根	答案
明天	日	月	一	大	JEGD
发现	乙	丿	王	冂	NTGM

二、三字词组的录入

三字词组的输入方法为：

首字第一个字根＋次字第一个字根＋第三个字的前两个字根

示例见表 2—10—2。

表 2—10—2 三字词组

例词	首字 第一个字根	次字 第一个字根	第三个字 第一个字根	第三个字 第二个字根	答案
计算机	讠	竹	木	几	YTSM
服务员	月	夂	口	贝	ETKM

三、四字词组的录入

四字词组的输入方法为：

按顺序各取每个字的第一个字根

示例见表 2—10—3。

表 2—10—3 四字词组

例词	首字 第一个字根	次字 第一个字根	第三个字 第一个字根	第四个字 第一个字根	答案
民主党派	乙	丶	⺌	氵	NYII
科学技术	禾	⺍	扌	木	TIRS

四、多字词组的录入

多字词组的输入方法为：

前三个字的第一个字根＋最后一个字的第一字根

示例见表 2—10—4。

表 2—10—4　　多字词组

例词	首字 第一个字根	次字 第一个字根	第三个字 第一个字根	最后一字 第一个字根	答案
中央电视台	口	冂	日	厶	KMJC
中华人民共和国	口	亻	人	囗	KWWL

修行靠个人

要求：在上机练习前先写出下列词组的编码，然后通过上机练习加以验证，之后再通过反复的上机练习（10～20 遍）达到巩固和熟练掌握的目的。

一、二字词组的练习

目标：录入速度 60 字/分钟，即在 5 分钟以内完成录入。

只要（　　　）　一起（　　　）　以为（　　　）　先生（　　　）
然后（　　　）　回来（　　　）　管理（　　　）　并不（　　　）
原来（　　　）　必须（　　　）　那些（　　　）　表示（　　　）
应该（　　　）　无法（　　　）　许多（　　　）　任何（　　　）
情况（　　　）　男人（　　　）　需要（　　　）　明白（　　　）
不可（　　　）　因此（　　　）　改革（　　　）　事情（　　　）
国际（　　　）　相信（　　　）　得到（　　　）　所有（　　　）
朋友（　　　）　喜欢（　　　）　下去（　　　）　似乎（　　　）
要求（　　　）　北京（　　　）　出现（　　　）　完全（　　　）
说话（　　　）　通过（　　　）　看见（　　　）　我国（　　　）
发生（　　　）　研究（　　　）　过来（　　　）　公司（　　　）
离开（　　　）　美国（　　　）　对方（　　　）　真是（　　　）
其中（　　　）　大家（　　　）　方面（　　　）　领导（　　　）
出去（　　　）　决定（　　　）　这时（　　　）　信息（　　　）
组织（　　　）　实现（　　　）　有关（　　　）　心里（　　　）
不同（　　　）　活动（　　　）　合作（　　　）　成为（　　　）
根本（　　　）　一般（　　　）　继续（　　　）　特别（　　　）
服务（　　　）　于是（　　　）　正是（　　　）　教育（　　　）
解决（　　　）　感觉（　　　）　精神（　　　）　非常（　　　）
方法（　　　）　父亲（　　　）　提高（　　　）　以后（　　　）

她们（ ） 由于（ ） 提出（ ） 目前（ ）
几乎（ ） 一面（ ） 代表（ ） 之中（ ）
文化（ ） 其他（ ） 主要（ ） 思想（ ）
母亲（ ） 会议（ ） 回答（ ） 生产（ ）
清楚（ ） 部门（ ） 群众（ ） 小姐（ ）
也许（ ） 加强（ ） 兄弟（ ） 甚至（ ）
难道（ ） 作为（ ） 建立（ ） 自然（ ）
结果（ ） 实在（ ） 经过（ ） 环境（ ）
处理（ ） 最后（ ） 人员（ ） 机会（ ）
对于（ ） 人家（ ） 不错（ ） 科技（ ）
其实（ ） 历史（ ） 能够（ ） 电话（ ）
立即（ ） 认识（ ） 中央（ ） 工程（ ）
多少（ ） 了解（ ） 然而（ ） 地区（ ）
参加（ ） 意思（ ） 不断（ ） 女儿（ ）
同志（ ） 看来（ ） 支持（ ） 提供（ ）
政治（ ） 准备（ ） 数据（ ） 整个（ ）
开发（ ） 结构（ ） 那样（ ） 发出（ ）
努力（ ） 农民（ ）

二、三字词组的练习

目标：录入速度60字/分钟，即在5分钟以内完成录入。

幼儿园（ ） 太阳能（ ） 出版社（ ） 国务院（ ）
事实上（ ） 计算机（ ） 数据库（ ） 解放军（ ）
大学生（ ） 复印机（ ） 党中央（ ） 自行车（ ）
共青团（ ） 小朋友（ ） 电视台（ ） 教育部（ ）
铁道部（ ） 发言人（ ） 办事员（ ） 国防部（ ）
世界观（ ） 私有制（ ） 人民币（ ） 普通话（ ）
运动员（ ） 气象台（ ） 研究所（ ） 中纪委（ ）
国庆节（ ） 劳动节（ ） 党支部（ ） 文化馆（ ）
目的地（ ） 中学生（ ） 总书记（ ） 代表团（ ）
公安部（ ） 同志们（ ） 留学生（ ） 办公室（ ）
大规模（ ） 科学院（ ） 总工会（ ） 洗衣机（ ）
科学家（ ） 共和国（ ） 私有制（ ） 全世界（ ）

基本上（ ） 洗衣机（ ） 介绍信（ ） 共产党（ ）
各单位（ ） 组织部（ ） 大使馆（ ） 上海市（ ）
西安市（ ） 河北省（ ） 吉林省（ ） 联合国（ ）
天安门（ ） 教育部（ ） 出版社（ ） 文汇报（ ）
阅览室（ ） 代表团（ ） 年轻化（ ） 无线电（ ）
所有制（ ） 动物园（ ） 水电站（ ） 目的地（ ）
同志们（ ） 中学生（ ） 发动机（ ） 办公室（ ）
基本上（ ） 领事馆（ ） 系列化（ ） 专业户（ ）
主动性（ ） 可行性（ ） 产品税（ ） 电视台（ ）
行政区（ ） 大规模（ ） 科学院（ ） 总工会（ ）
科学家（ ） 共和国（ ） 私有制（ ） 全世界（ ）
基本上（ ） 洗衣机（ ） 介绍信（ ） 共产党（ ）
共和国（ ） 私有制（ ） 全世界（ ） 组织部（ ）

三、四字词组的练习

目标：录入速度 60 字/分钟，即在 5 分钟以内完成录入。

共产党员（ ） 奋发图强（ ） 党政机关（ ）
基本原则（ ） 同甘共苦（ ） 基本原则（ ）
水落石出（ ） 少数民族（ ） 形式主义（ ）
中国人民（ ） 社会主义（ ） 千方百计（ ）
全党全国（ ） 无产阶级（ ） 中华民族（ ）
知识分子（ ） 中国政府（ ） 新华书店（ ）
天气预报（ ） 中央委员（ ） 生活水平（ ）
劳动模范（ ） 计划生育（ ） 共产主义（ ）
全国各地（ ） 众所周知（ ） 先进集体（ ）
家喻户晓（ ） 全心全意（ ） 少先队员（ ）
内部矛盾（ ） 唯物主义（ ） 自始至终（ ）
群众路线（ ） 社会科学（ ） 体力劳动（ ）
自动控制（ ） 行政管理（ ） 思想方法（ ）
振兴中华（ ） 炎黄子孙（ ） 科学技术（ ）
唯心主义（ ） 机构改革（ ） 高等院校（ ）
文明礼貌（ ） 社会实践（ ） 民主党派（ ）
海外侨胞（ ） 人民政府（ ） 工人阶级（ ）

爱国主义（　　　　） 精神文明（　　　　） 中国青年（　　　　）
新陈代谢（　　　　） 大公无私（　　　　） 艰苦奋斗（　　　　）
科学分析（　　　　） 国民经济（　　　　） 国际机场（　　　　）
个人利益（　　　　） 标点符号（　　　　） 公共汽车（　　　　）
科研成果（　　　　） 生动活泼（　　　　） 中国银行（　　　　）
自力更生（　　　　） 精兵简政（　　　　） 后来居上（　　　　）
后顾之忧（　　　　） 名胜古迹（　　　　） 人尽其才（　　　　）
精益求精（　　　　） 百家争鸣（　　　　） 见义勇为（　　　　）

四、多字词组的练习

目标：录入速度60字/分钟，即在4分钟以内完成录入。

中央电视台（　　　　） 中国共产党（　　　　） 中央委员会（　　　　）
中央办公厅（　　　　） 常务委员会（　　　　） 全民所有制（　　　　）
人民大会堂（　　　　） 中央政治局（　　　　） 军事委员会（　　　　）
喜马拉雅山（　　　　） 西藏自治区（　　　　） 民主集中制（　　　　）
毛泽东思想（　　　　） 中国科学院（　　　　） 发展中国家（　　　　）
中央电视台（　　　　） 中国共产党（　　　　） 中央委员会（　　　　）
历史唯物主义（　　　　） 内蒙古自治区（　　　　）
中国人民解放军（　　　　） 政治协商会议（　　　　）
中华人民共和国（　　　　） 中央人民广播电台（　　　　）
广西壮族自治区（　　　　） 新疆维吾尔自治区（　　　　）
中央书记处（　　　　） 全国人民代表大会（　　　　）
历史唯物主义（　　　　） 内蒙古自治区（　　　　）
中国人民解放军（　　　　） 政治协商会议（　　　　）
中华人民共和国（　　　　） 中央人民广播电台（　　　　）
广西壮族自治区（　　　　） 新疆维吾尔自治区（　　　　）
中央书记处（　　　　） 全国人民代表大会（　　　　）
中国人民解放军（　　　　） 中华人民共和国（　　　　）

灵丹妙药

一、圈圈点点

从报纸或杂志中任意找一篇文章，把上面的词组圈出来，然后逐一校对检查。如此反复多次，可以提高自己对于常用词组的敏感度，最终达到能“下意识”地进行输入的程度。

二、注重积累

语文中的词组在五笔词库中不一定都是词组，在日常录入过程中要注重积累。对于已经确定是五笔字库中的词组，要专门收集并进行集中练习。例如可以将收集到的词组编写成小文章，进行练习。

三、强行记忆

通过长时间的词组收集，你会发现五笔字库中部分词组在日常的各类文章中出现的频率非常高，例如我们、中国、没有、可以等，对于这类词要强记，形成条件反射，对于提高日常文章的录入速度会很有帮助。

过关斩将

要求：反复练习以下每部分的测试内容，测试时间为5分钟，必须达到60字/分钟，才能进入下一部分的测试，全部测试完成后可以进入活动12。

第一关：词组录入（二、三、四、多字词组各占25%。）

过去　过程　自然　城市　这么　文章　真正　根据　后来　学习
进入　人员　提出　中心　无法　老师　达到　方法　女人　内容
事情　是否　决定　可是信息　印度洋　成都市　幼儿园
日用品　专利法　年轻人　展销会　杭州市　人民币
国务院　数据库　事实上　联合国　世界观　教研室
消费品　责任制　自行车　标准化　天安门　医学院
招待所　大部分　灵敏度　专业化　基本原则　同甘共苦
克勤克俭　光彩夺目　水落石出　夜以继日　少数民族　政协委员
形式主义　内部矛盾　中国人民　社会主义　劳动模范　经济基础
计划生育　共产主义　工作人员　劳动人民　人民日报　科学研究
全心全意　自力更生　银行账号　群众路线　衣食住行
新疆维吾尔自治区　常务委员会　中国人民解放军
全国人民代表大会　民主集中制　毛泽东思想　人民大会堂
中华人民共和国　宁夏回族自治区　新技术革命　西藏自治区
新华通讯社　四个现代化　发展中国家　宁夏回族自治区
军事委员会　为人民服务　集体所有制　全国人民代表大会
政治协商会议　中国人民解放军

第二关：词组录入（二字词组占40%，三、四、多字词组各占20%。）

否认 调查 放松 姑娘 奋斗 动态 多种 号召
谨慎 合作 合同 钢笔 规定 对方 服装 单位
导演 出生 保持 玻璃 加工 假如 钢铁 介绍
季节 监狱 措施 结合 家具 汉族 检查 登记
广告 错误 答案 东方 关键 道路 电视 代理
共同 挫折 程度 惭愧 出差 彩色 测量 村庄
操作 常委 唱歌 春节 处境 出差 储蓄 车辆
处分 籍贯 合适 集团 工艺品 规律性 发动机
科学家 文化馆 台北市 机械化 中纪委 出版社 展销会
发电机 办公室 四川省 共产党 介绍信 科学家 后勤部
全世界 中学生 目的地
爱国主义 分发图强 党政机关 个人利益 海外侨胞 国防大学
银行账号 总结经验 众所周知 高等院校 人民政府 程序设计
新陈代谢 天气预报 中共中央
新疆维吾尔自治区 全国人民代表大会 常务委员会 人民代表大会
中国共产党 中央办公厅 中央电视台 中央书记处 中央委员会
宁夏回族自治区 为人民服务 中央政治局 广西壮族自治区
民主集中制 新技术革命 新华通讯社 喜马拉雅山 军事委员会
中国科学院 发展中国家 集体所有制 四个现代化

第三关：词组录入（二字词组占70%，三、四字词组各占15%。）

目前 一直 记者 地方 希望 部分 上海 发生 能力
必须 以后 这里 怎么 精神 结果 选择 一点 方式
之间 喜欢 不要 其实 能够 或者 得到 任何 思想
那些 当然 存在 包括 完全 一起 活动 使用 一般
专业 国际 提供 然后 有关 了解 政治 如何 原因
传统 一切 并不 朋友 管理 于是 具有 系统 如此
产品 计划 那个 告诉 过去 过程 自然 城市 这么
文章 真正 根据 后来 学习 进入 人员 提出 中心
无法 老师 达到 方法 女人 内容 事情 是否 决定
复杂 房屋 户口 公益 调查 季节 记号 道路 高峰
改进 敌人 发明 观众 规则 放大 分别 固体 高兴
符号 锻炼 繁荣 粉碎 测验 可爱

普通话　消费品　领事馆　阅览室　计算机　服务员　共和国　摩托车　党支部　可行性　全世界　上海市　消费品　教育部　发言人
家喻户晓　生活水平　新华书店　炎黄子孙　全心全意　国家机关
经济效益　交通规则　科研成果　一分为二　千方百计　银行账号

活动12　融会贯通——录入文章

一大早，小文和小璐就在师傅的书房里急切地等待师傅的到来。

“今天怎么来得这么早啊?”师傅大声问道。

“师傅，上次跟您说过的我们想参加中英文录入争霸赛，您看……，再说您是一代名师，我们想通过我们参加比赛，也给师傅您露露脸呀!”

“这样吧，师傅今天先问问你们，文章中的标点符号你们进行过专门练习吗?”

“这还没有?”小文和小璐异口同声地回答道。

“那你们还注意到自己训练中还存在哪些问题吗，需要师傅指导的。”

“暂时还没有发现。”小文和小璐回答说。

“两位小徒弟，赛场上比赛，高手过招，失之毫厘，差之千里！训练中的任何细节都要注意，你在赛场上才不至于差之千里。另外，即便你有高的水平，也不能自傲，要能不断发现自己的不足之处，并予以改进，这样才能立于不败之地，否则，即便侥幸成功，也不会长久。”

小文和小璐不好意思地低下了头。

“师傅今天就针对文章中的标点符号练习帮你们略微指点一二。标点符号的练习也一定要注意准确性和速度，也要实现“盲打!”。此外，文章录入的过程中也一定要注意使用简码和词组。”

小璐说：“谢谢师傅的教导，我们一定戒骄戒躁，注意文字录入中的细节，请师傅放心!”

小文也大声附和：“请师傅放心!”

摩拳擦掌

中国的语言文字丰富多彩，在精彩的文字中，标点符号是不可缺少的组成部分，文章中常用的标点符号（见表2—11—1）。

表 2—11—1　　中文常用标点符号

名称	符号	名称	符号
逗号	，	省略号	……
顿号	、	句号	。
分号	；	破折号	——
冒号	：	感叹号	！
双引号	“”	书名号	《》
单引号	‘’	括号	()

师傅领进门

一、标点符号的输入方法，见表 2—11—2

表 2—11—2　　中文常用标点符号输入

名称	符号	对应键位	名称	符号	对应键位
逗号	，	，	省略号	……	ˆ
顿号	、	/或\	句号	。	.
分号	；	；	破折号	——	—（上档键）
冒号	：	：	感叹号	！	！
双引号	“”	“”	书名号	《》	<>
单引号	‘’	‘’	括号	()	()

注意：有些键上有上、下两个键，这时要借助【Shift】键，即用右手打特殊符号时，左手小指按住【Shift】键，若用左手打特殊符号，则用右手小指按住【Shift】键。

二、文章中标点符号的输入也一定要实现“盲打”!

标点符号是在任何文章录入过程中都会碰到的，如果标点符号录入没有实现盲打或者打错，一定会影响整体的录入速度。

三、练习文章的类型要广泛

不同文体的文章录入时有各自的特点，例如政论类型文章词组较多，能更好地练习词组的拆分方法和掌握拆分原则；小说、散文类型的文章标点符号丰富，除了练习拆字外，还有助于更好地练习标点符号；古文则有助于提升单字拆分的能力。

四、文章练习要有深度

这里的深度是指对所练习过的文章要练熟、练透。每篇文章的练习都不能一遍而过，一般要经过 5～10 遍的反复操练。第一遍录入完成后，快速将卡壳和错字以及新发现的词组进行登记，专门练习，练熟后再开始第二遍，依此类推，直到录入非常顺畅为止，再开始新文章的练习。

修行靠个人

要求：反复练习以下每部分的输入内容，直到该部分的目标达成以后，才能进入下一个部分的输入练习。

一、标点符号练习

目标：录入速度达到 60 字符/分钟，即在 3 分钟以内完成录入。

，、；：“”‘’——　。……　！？《》＜＞（），、；：“”‘’——　。……　！？《》＜＞（），、；：“”‘’——　。……　！？《》＜＞（），、；：“”‘’——　。……　！？《》＜＞（），、；：“”‘’——　。……　！？《》＜＞（），、；：“”‘’——　。……　！？《》＜＞（），、；：“”‘’——　。……　！？《》＜＞（），、；：“”‘’——　。……　！？《》＜＞（），、；：“”‘’——　。……　！？《》＜＞（）—　。……　！？

二、不同类型文章练习

1. 输入以下政论类型文章

目标：录入速度达到 60 字符/分钟，以 5 分钟为单位进行练习。

青年最富有朝气、最富有梦想。近代以来，我国青年不懈追求的美好梦想，始终与振兴中华的历史进程紧密相连。在革命战争年代，广大青年满怀革命理想，为争取民族独立、人民解放冲锋陷阵、抛洒热血；在社会主义革命和建设时期，广大青年响应党的号召，向困难进军，向荒原进军，保卫祖国，建设祖国，在新中国的广阔天地忘我劳动、艰苦创业；在改革开放历史新时期，广大青年发出团结起来、振兴中华的时代强音，为祖国繁荣富强开拓奋进、锐意创新。在最近的芦山抗震救灾中，大批青年临危不惧、顽强拼搏，广大青年心系灾区、无私奉献，为抗震救灾做出了重要贡献。

历史和现实都告诉我们，青年一代有理想、有担当，国家就有前途，民族就有希望，实现我们的发展目标就有源源不断的强大力量。

在革命、建设、改革各个历史时期，中国共产党始终高度重视青年、关怀青年、信任青年，对青年一代寄予殷切期望。

2. 输入以下散文类文章

目标：录入速度达到60字符/分钟，以5分钟为单位进行练习。

于是我又回忆起另一个画面，这就在所谓“黄土高原”！那边山多数是秃顶的，然而层层的梯田，将秃顶装扮成稀稀落落有些黄毛的癞头，特别是那些秆植物颀长而整齐，像等待检阅的队伍似的，在晚风中摇曳，别有一种惹人怜爱的姿态。可是更妙的是三五月明之夜，天是那样的蓝，几乎透明似的，月亮离山顶似乎不过几尺，远看山顶的谷子丛密挺立，宛如人头上的怒发。这时候忽然从山脊上长出两支牛角来，随即牛的全身也出现，掮着犁的人形也出现，并不多，只有两个，也许还跟着小孩，他们姗姗而下，在蓝的天，黑的山，银色的月光的背景上，成就了一幅剪影，如果给田园诗人见了，必将赞叹为绝妙的题材。可是没有完。这几位晚归的种地人，还把他们那粗朴的短歌，用愉快的旋律，从山顶上飘下来，直到他们没入山坳，依旧只有蓝天明月黑的山，歌声可是缭绕不散。

3. 输入以下文言文

目标：录入速度达到30字符/分钟，以5分钟为单位进行练习。

故今日之责任，不在他人，而全在少年。少年智则国智，少年富则国富，少年强则国强，少年独立则国独立，少年自由则国自由，少年进步则国进步，少年胜于欧洲，则国胜于欧洲，少年雄于地球，则国雄于地球。红日初升，其道大光。河出伏流，一泻汪洋。潜龙腾渊，鳞爪飞扬。乳虎啸谷，百兽震惶。鹰隼试翼，风尘翕张。奇花初胎，矞矞皇皇。干将发硎，有作其芒。

灵丹妙药

一、善于找差补漏

文章录入过程也是寻找自身录入短板的过程，要注重收集错字、难拆字和新词组。对收集到的字和词组进行专门、反复的练习。

二、注重简码词组

简码和词组是提升文章录入速度的“法宝”。使用简码和词组，能降低码长，提升单位时间内的有效键数。

三、巧用练习软件

文章录入阶段建议使用有统计功能的、能反映成绩增长轨迹的练习软件，例如打字高手、打字旋风等。图 2—12—1 所示为打字高手软件，能清晰反映成绩增长曲线。

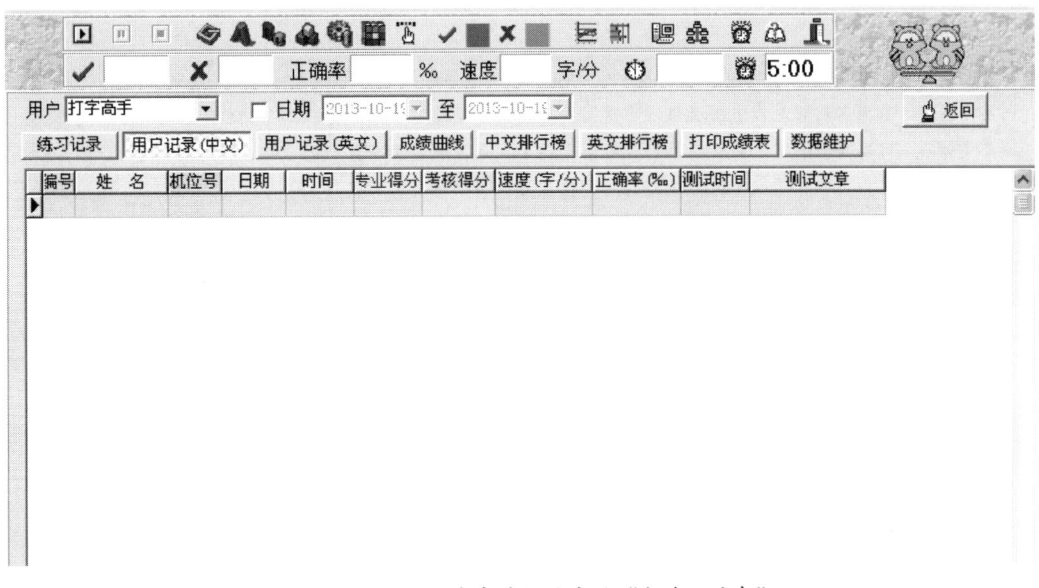

图 2—12—1 打字高手软件中的“成绩统计表”

过关斩将

要求：以下测试内容，以 10 分钟为一个测试单元，达到 60 字/分钟，才能进入下一个环节的测试，全部测试完成后可以进入项目三部分的学习。

第一关：政论文章录入练习

文章内容	累计字数
中国梦是实干复兴之梦。自古以来，老百姓不仅看当权者“怎么说”，	31
更看其“如何做”。我们党能否继续赢得人民，关键还在一个“干”字而	64
且是务实地干、进取地干。中国共产党追梦、圆梦的 90 多年，就是务实	96
进取的90 多年。历史证明：在多少次遭遇危险和困难时，无不都是党的	126
务实力挽狂澜，使中国梦涅槃重生。例如，党从国情出发选择“农村包围	154
城市”，有效应对了 1927 年大革命失败后面临的极大危险；党适应形势开	186
展长征，有效应对了 1934 年中央根据地反“围剿”失败后的严重困难；	212
党结合实际全面整顿，有效应对了 1959 年至 1961 年的严重经济困难；党	239
实事求是拨乱反正，有效应对了“文化大革命”造成的严峻局面；党解放	265
思想、改革开放，有效应对了全球化条件下再次“落后挨打”的潜在风险，	292
等等。务实则党兴国兴、好梦成真，不务实则误党误国、美梦破碎。当下，	319
“务实”二字，鲜明地道出了中国梦的实践路径和基本方法，解决了中国	347
梦“怎么追”、“怎么圆”的问题。这就要求我们在实现中国梦的过程中，	372

文章内容	累计字数
必须牢牢把握当代中国发展的总依据、总布局、总任务和“三个没有变”	400
的基本国情，实事求是地走好中国道路，确保实现中国梦的正确方向，做	429
到不走错路、邪路和回头路；必须实事求是地推进改革发展，以实干凝聚	458
中国力量，确保中国梦有得以实现的强大物质文化基础；必须坚决与形式	488
主义、官僚主义、享乐主义和奢靡之风决裂，以唯实之态形成中国风气，	517
确保在实现中国梦的历程中少一些干扰、少一些徘徊、少一些折腾。党的	546
群众路线教育实践活动把“务实”作为三大主题之一，使中国梦有了最质	576
朴、最能动也最有效的方法论，形成了“人人都是实干家”的热烈氛围。	603
既高瞻远瞩又脚踏实地，我们党带领全国人民走过了“雄关漫道真如铁”	632
的昨天，迎来了“人间正道是沧桑”的今天，顺着这条务实之路，中国梦	670
必将会在“长风破浪会有时”的美好明天得以实现。	690

第二关：散文录入练习

文章内容	累计字数
从火车上遥望泰山，几十年来有好些次了，每次想起“孔子登东	25
山而小鲁，登泰山而小天下”那句话来，就觉得过而不登，像是欠下	52
悠久的文化传统一笔债似的。杜甫的愿望：“会当凌绝顶，一览众山小”，	78
我也一样有，惜乎来去匆匆，每次都当面错过了。而今确实要登泰山	105
了，偏偏天公不作美，下起雨来，淅淅沥沥，不像落在地上，倒像落	130
在心里。天是灰的，心是沉的。我们约好了清晨出发，人齐了，雨却	155
越下越大。等天晴吗？想着这渺茫的等字，先是憋闷。盼到十一点半	181
钟，天色转白，我不由喊了一句：“走吧！”带动年轻人，挎起背包，兴	205
致勃勃，朝岱宗坊出发了。	215
是烟是雾，我们辨认不清，只见灰蒙蒙一片，把老大一座高山，	239
上上下下，裹了一个严实。古老的泰山越发显得崔嵬了。我们才过岱	266
宗坊，震天的吼声就把我们吸引到虎山水库的大坝前面。七股大小，	293
从水库的桥孔跃出，仿佛七幅闪光黄锦，直铺下去，碰着嶙嶙的乱石，	313
激起一片雪白水珠，脱线一般，撒在洄旋的水面。这里叫做虬在湾：	339
据说虬早已被吕洞宾度上天了，可是望过去，跳掷翻腾，像又回到了	366
故居。我们绕过虎山，站到坝桥上，一边是平静的湖水，迎着斜风细	392
雨，懒洋洋只是欲步不前，一边喑哑叱咤，似有千军万马，躲在绮丽	418
的黄锦底下。黄锦是方便的比喻，其实是一幅细纱，护着一幅没有经	445
纬的精致图案，透明的白纱轻轻压着透明的米黄花纹。——也许只有织女	479
才能织出这种瑰奇的景色。雨大起来了，我们拐进王母庙后的七真祠。	506
这里供奉着七尊塑像，正面当中是吕洞宾，两旁是他的朋友李铁拐和	534
何仙姑，东西两侧是他的四个弟子，所以叫做七真祠。吕洞宾和他的	561

文章内容	累计字数
两位朋友倒也罢了，站在龛里的两个小童和柳树精对面的老人，实在	590
是少见的传神之作。一般庙宇的塑像，往往不是平板，就是怪诞，造	616
型偶尔美的，又不像中国人，跟不上这位老人这样逼真、亲切。无名	641
的雕塑家对年龄和面貌的差异有很深的认识，形象才会这栩栩如生。	667
不是年轻人提醒我该走了，我还会欣赏下去的。	686

第三关：混合文本录入练习

文章内容	累计字数
那是力争上游的一种树，笔直的干，笔直的枝。它的干呢，通常	24
是丈把高，像是加以人工似的，一丈以内，绝无旁枝；它所有的桠枝	51
呢，一律向上，而且紧紧靠拢，也像是加以人工似的，成为一束，绝	76
无横斜逸出；它的宽大的叶子也是片片向上，几乎没有斜生的，更不	103
用说倒垂了；它的皮，光滑而有银色的晕圈，微微泛出淡青色。这是	129
虽在北方的风雪的压迫下却保持着倔强挺立的一种树！哪怕只碗来粗	158
细罢，它却努力向上发展，高到丈许，两丈，参天耸立，不折不挠，	182
对抗着西北风。这就是白杨树，西北极普通的一种树，然而绝不是平凡	209
的树！	212
莫知吾所以制胜之形。故其战胜不复，而应形于无穷。夫兵形象	239
水，水之行，避高而趋下；兵之形，避实而击虚。水因地而制流，兵因敌	251
而制胜。故兵无常势，水无恒形。能因敌变化而取胜者，谓之神。故	277
五行无常胜，四时无常位，日有短长，月有死生。	295
一个老奶奶很讲究忌讳，逢年过节她总是吉利话不离口，从没说	318
过一个“不”字。一次大年初一，老奶奶一起床，小孙女就送来一碗	343
甜黏粥，她高兴地喝了。孙女问：“奶奶，再喝一碗好吗?”老奶奶回	366
答:“好,好。”小孙女立刻送来第二碗黏粥，她又喝了。小孙女问：“再	390
来一碗?”老奶奶想到年节不能说“不”字，于是说：“好吧，我能喝	412
三碗。”就这样老奶奶一气喝了六碗，她的肚皮被撑得像一面大鼓。不	439
懂事的小孙女仍一个劲地问：“奶奶，你可愿意再喝一碗?”老奶奶不	465
由自主地连忙摇手说:“不，不喝了，再喝一点儿，奶奶就要胀死了啊!”	491
一座“绿色工厂”开工了，在太阳光的照射下，“机器”开动，原	513
料是从根部送来的水分，加上通过叶面许许多多开闭自如的气孔从空	540
气中“收购”来的二氧化碳，依靠神通广大的“化学工程师”酶的密	564
切配合，经过一系列奇妙复杂的变化，制成了碳水化合物。如果再加	591
上土壤里取得的氮、硫、磷等无机元素，就可以进一步造出脂肪、蛋	615
白质等物质来。	621

藏经阁

一、五笔字型输入编码歌诀

五笔字型均直观，依照笔顺把码编
键名汉字打四下，基本字根请照搬
一二三末取四码，顺序拆分大优先
不足四码要注意，交叉识别补后边

二、五笔字型汉字编码流程图

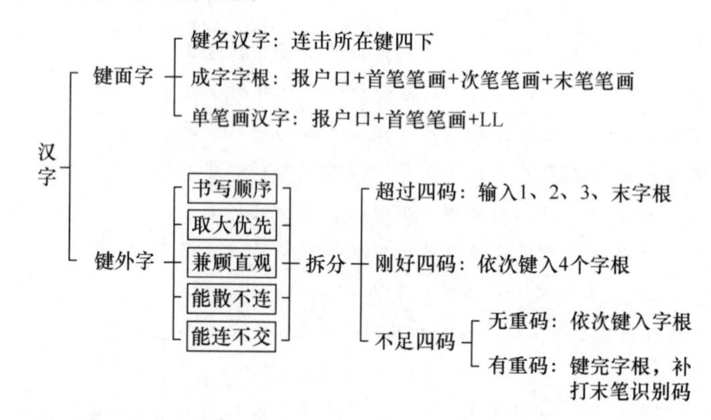

三、五笔词组录入方法汇总

词组类别	录入方法
二字词	首字的前两个字根＋次字的前两个字根
三字词	首字第一个字根＋次字第一个字根＋第三个字的前两个字根
四字词	按顺序各取每个字的第一个字根
多字词	前三个字的第一个字根＋最后一个字的第一字根

打擂台

以下中文文稿需要录入，记录完成的时间，并参照下表给出自我评价。中级水平的录入速度 80 字符/分钟是每个人通过短期努力都可以达到的。

中文录入自我评价表

级别	完成时间（分钟）	录入速度（字/分钟）	称号	文字录入四级鉴定标准
入门	28	30	打字新手	—
初级	17	50	打字老手	合格
中、低级	10.5	80	打字熟手	良好
中级	8.5	100	打字高手	优秀
中、高级	7	120	打字健将	—
高级	5	170	打字达人	—

测试文章

文章内容	累计字数
雷锋，原名雷正兴，1940年出生在湖南省望城县一个贫苦农家。	27
雷锋生前是解放军沈阳部队工程兵某部运输班班长、五好战士，1962	56
年8月15日因公殉职。他的爱憎分明、言行一致、公而忘私、奋不	85
顾身、艰苦奋斗、助人为乐，把有限的生命投入到无限的为人民服	114
务之中去的崇高精神，集中体现了中华民族的传统美德和共产主义	143
道德品质。雷锋入伍后，他被编入工程兵某部运输连四班当汽车兵。	173
1960年11月，他加入了中国共产党。他入伍后表现突出，沈阳军	200
区《前线报》开辟了“向雷锋学习”的专栏。在不到三年的时间里，	229
他荣立二等功一次、三等功两次，被评为节约标兵，荣获“模范共	259
青团员”,出席过沈阳部队共青团代表会议。1961年，雷锋晋升为	287
班长，被选为抚顺市人民代表。1962年8月15日，他因事故不幸	314
殉职。1963年3月5日，毛泽东主席为雷锋的题词“向雷锋同志学	342
习”在《人民日报》发表。此后，全国广泛开展学习雷锋的活动。	371
螺丝钉精神的由来在望城的山间小道上，一颗小小的螺丝钉同	398
时映入了张书记和雷锋的眼帘。小雷锋蹦蹦跳跳，一脚踢飞了螺丝	427
钉。张书记却上前几步，弯腰捡起来，把螺丝钉上的灰擦干净，郑	456
重地交给雷锋：“留着，会有用处的。”就这样一弯腰，一句话，	485
一个老共产党员的言行竟然影响了一个年轻人的一生。在后来雷锋	514
写的日记中，螺丝钉被雷锋反复思索，终于形成了独特的“螺丝钉	543
精神”。1960年1月12日，雷锋写道：“虽然是细小的螺丝钉，	570
是个细微的小齿轮，然而如果缺了它，那整个机器就无法运转了，	600
别说是缺了它，即使是一颗小螺丝钉没拧紧，一个小齿轮略有破损，	630
也要使机器的运转发生故障的。尽管如此，但是再好的螺丝钉，再	659
精密的齿轮，它若离开了机器这个整体，也不免要当做废料，扔到	688
废铁料仓库里去的。”1962年4月7日，雷锋再次写道：“一个人	716
的作用对于革命事业来说，就如一架机器上的一颗螺丝钉。机器由	745
于有许许多多螺丝钉的连接和固定，才成了一个坚实的整体，才能	774
运转自如，发挥它巨大的工作能力，螺丝钉虽小，其作用是不可估	803
量的，我愿永远做一颗螺丝钉。螺丝钉要经常保养和清洗才不会生	832
锈。人的思想经常检查才不会出毛病。”	850

IANGMUSAN

项目三 中英文混合录入

活动1　随机应变——状态切换

学完了前面两个项目，对于绝大部分的汉字，小文和小璐都能熟练、正确地输入，俩人可真是松了一口气。“只要再加把劲练练，我们就成为打字高手喽！”小文伸伸懒腰，得意地对小璐说。

“谁说光中文输入没问题就是打字高手了？喏，给你们一篇文章打打看！”不知什么时候师傅走了进来。

“没问题！师傅请出招！”凭着这段时间刻苦训练的功底，小文和小璐信心满满。

“就是这篇，试试吧。”师傅淡定地发出一篇中英数混合录入的文章。

不就是又有中文、又有英文、又有数字嘛！最难的中文录入都掌握了，这有何难?！深呼吸，调整好坐姿之后，小文和小璐开始录入了。一开始录中文还好，后面一会中文一会英文一会数字，又输错，再按回退键进行修改，俩人手忙脚乱了半天，好不容易打完了10分钟，已经是满头大汗。再看成绩，惨啊！连30字/分钟都不到！

“师傅，这怎么回事啊?”小文和小璐满脸困惑。

“这就是我今天要教你们的状态切换！”师傅微微一笑。

摩拳擦掌

主键盘区有左右对称的三个键：【Ctrl】【Shift】【Alt】，如图3—1—1所示。

图3—1—1　切换键示意图

这三个键和其他键组合可以进行不同输入法、中英文标点符号以及全角和半角等状态的切换。

师傅领进门

一、中英文输入法切换：【Ctrl】＋【Space】

指法：左手或右手的小指按住【Ctrl】键，左手或右手的拇指轻点【Space】键。

（具体使用左手还是右手根据个人的习惯，舒服就好。）

二、不同输入法切换：【Ctrl】+【Shift】

指法：左手的小指按住左边的【Ctrl】键，右手的小指轻点右边的【Shift】键。

三、中英文标点符号切换：【Ctrl】+【.】

指法：左手的小指按住【Ctrl】键，右手的无名指轻点【.】键。

四、全角/半角切换：【Shift】+【Space】

指法：左手或右手的小指按住【Shift】键，左手或右手的拇指轻点【Space】键。

修行靠个人

要求：反复练习以下每部分的输入内容，直到该部分的目标达成以后，才能进入下一部分的输入练习。

一、中英文输入法切换练习

1. 输入以下中英文混合文章（基础训练：中文一句、英文一句混合文章，中英文平均。）

目标：录入速度达到45字符/分钟，即在6分钟内完成录入。

今天我们拥有了更高层的楼宇以及更宽阔的公路，但是我们的性情却更为急躁，眼光也更加狭隘；Today we have higher buildings and wider highways，but shorter temperaments and narrower points of view；我们的住房更大了，但我们的家庭却更小了；We have bigger houses，but smaller families；我们拥有了更多的知识，可判断力却更差了；We have more knowledge，but less judgment；我们可以往返月球，但却难以迈出一步去亲近我们的左邻右舍。We reached the Moon and came back，but we find it troublesome to cross our own street and meet our neighbors.

2. 输入以下中英文混合文章（强化练习：5个中文字一个英文单词，练习切换。）

目标：录入速度达到35字符/分钟，即在6分钟内完成录入。

奇怪strange得很，这次却but有迥乎不同的印象impression。六月June，并不是好时候time，没有春光，没有雪snow，也没有秋Autumn意。那几天，有的是满湖烟雨misty rain，山光水water色，俱是一片piece迷蒙。西湖，仿佛as if在半醒半睡。空气air中，弥漫permeate着经了雨的栀子花gardenias的甜香。记起东坡诗句verse：“水光潋滟晴sunny day方好，山色空蒙雨rain亦奇。”便想，东坡自是最了解under-stand西湖的人，实在应该should仔细观赏，领略appreciate才是。

3. 输入以下中英文混合文章（综合练习：中文、英文混合文章，中文多、英文少。）

目标：录入速度达到30字符/分钟，即在9分钟内完成录入。

正像每次一样，匆匆地来，又匆匆地去。几天中我领略了两个字，一个是“绿”，只凭这一点，已使我流连忘返。雨中去访灵隐 Lingyin Temple，一下车，只觉得绿意扑眼而来。道旁古木 furuki 参天，苍翠欲滴，似乎飘着的雨丝儿也都是绿的。飞来峰 Klippes 上层层叠叠的树木，有的绿得发黑，深极了，浓极了；有的绿得发蓝，浅极了，亮极了。峰下蜿蜒的小径 alley，布满青苔，直绿到了石头缝里。在冷泉亭 Cold Spring Pavilion 上小坐，真觉得遍体生闲，心旷神怡。亭旁溪水，说是溪水，其实表达不出那奔流的气势，平稳处也是碧澄澄的，流得急了，水花飞溅，如飞珠滚玉一般，在这一片绿色的影中显得分外好看。

二、不同输入法切换练习

输入以下练习内容（要求一段换一种中文输入方法，可用的输入方法有五笔86版、智能ABC、微软拼音等。）

目标：录入速度达到30字/分钟，即在10分钟内完成录入。

随着信息社会的发展，打字已经不再局限于文稿的处理，而逐渐成为信息交流的一种手段。现在网络已成为人们交流的主要平台，上千字的文章可瞬间传输，文字交流已经成为人们交流的主要方式。

由于文字传输速度快，效率高，十分经济。一小时的通话内容以文字的形式几秒钟即可传输，节约费用上百倍，节约时间上千倍。如此大的吸引力，必然引发打字的革命。

每分钟几十字的录入速度显然不适应信息交流的需求。无论是个人或是单位，都面临着录入速度的竞争和挑战。

对信息资源的获取、传递、存储能力的大小和速度的快慢，在一定程度上决定了“竞争的胜负”。

这种“能力”，在一定程度上就是由“打字的能力”、“打字的速度”和“打字的准确性”所决定的。

三、中英文标点符号切换练习

1. 输入以下标点符号（基础训练：一行切换一次。）

目标：录入速度达到60字符/分钟，即在3分钟以内完成录入。

,;。、 ？ • ‘’、; “” ……——。,!:、。¥;:;《》

, \.@!:’ ”, .; \/? \”:’ \. “, ./! _; .?’ \.

? • ‘’,;:; “”;。、。,;:;《》……,。、 •、; ¥。

, .; \!:;’ “/., $/” .,:” \! .;’;” .!? .’ \

,! •。“;、,!、。;?;”。——……?。‘、!:。“

, . @:! _ \’, . ; \ ,;! _ “:?!,; . $!’: . →;”

2. 输入以下标点符号（强化练习：一组切换一次。）

目标：录入速度达到 60 字符/分钟，即在 3 分钟以内完成录入。

,;。、!:’ ”、; “” /? \” —— “, . /、 ¥ ,;?’ \ .

, \ . /? • ‘’, $; \ ……:’ \ . 。,!: ! _ ; . :; 《》

, . ; \ ,;:; “”;。“/. ,:; 《》,:” \ ,。、 • ? . $ @

? • ‘’!:;’、。,; \ /” . ……! . @’;” . !、;:。

, . /: “;、,, . ; \ ?;”。“:?! ¥ 。、:!’: . !:。“

\ ?;, ”! • 。! _ \ ’!、。;,;! _ ——,; . \ ?。‘、

四、全角/半角切换练习

1. 输入以下内容（基础训练：一行切换一次，第一行全角、第二行半角，依此类推。）

目标：录入速度达到 130 字符/分钟，即在 2 分钟以内完成录入。

ABUE OPQM QLSI PIDV ISDM UVWM

ABUE OPQM QLSI PIDV ISDM UVMW

HUSW LODE MUSN KLWR WEDV OKWX

HUSE LODE MUSM KLWR WEDV OKWX

DWGU KIAB LOPS JEBU OSMX QIDB

DWGU KIAB LOPS JEBU OSMX QIDB

GUPD XDLU OKSW CHEM LPXE QIDM

GUPD XDLU OKSW CHEM LPXE QIDM

2. 输入以下内容（强化练习：一组切换一次。）

目标：录入速度达到 130 字符/分钟，即在 2 分钟以内完成录入。

about love quit visit load move

them team live leaf your away

normal candy mother over father nother

sister mean pencial club teacher him

seem student brother hub switch her

rich swatch theaf time catch number

beam away duck high feet always

nothing victory have none moon home

3. 输入以下内容（综合练习：注意数字是用主键盘区上面一排数字进行录入。）

目标：录入速度达到 40 字符/分钟，即在 15 分钟以内完成录入。

凤凰科技讯北京时间 3 月 8 日消息，据国外科技博客 Business Insider 报道，当地时间周四，Facebook 在位于其加州门罗帕克的公司总部，向媒体发布了重新设计后的 News Feed（信息流）功能，这是七年来该公司对其主页进行的一次最大内容调整。改版后的 News Feed 设计内容包括放大了地图、文章、照片和应用信息（如 Pinterest 帖子）的图片。用户可选择按年代查看信息，也可像搜索音乐那样分类查看。

参加今天 Facebook 媒体发布会的除了该产品创始人之一、FacebookCEO 马克·扎克伯格（Mark Zuckerberg）外，还包括 News Feed 产品的另外两个发明人：Facebook 现任首席产品官——克里斯·考克斯（Chris Cox）和该产品的关键工程师——安德鲁·“波兹”·博斯沃思（Andrew“Boz”Bosworth）。

扎克伯格表示，Facebook 开发 News Feed 的目标是为用户创建一个“最个性化的报纸”，使他们能够了解朋友以及他们周围的世界发生了哪些最重要的事情。

Facebook 首席产品官克里斯·考克斯则表示，Facebook 打算为用户提供一个“更现代化、更简洁”的界面。Facebook 借鉴了智能手机和平板电脑的设计灵感，将其引入桌面 Web 版设计。考克斯暗示，Facebook 改进 News Feed 设计的目的只有一个，就是让用户将更多的时间花在 Facebook 上。

灵丹妙药

一、快速切换到五笔输入法状态

使用【Alt】键与其他键组合可以直接完成指定输入法切换，例如可以使用【Alt】＋【Shift】＋【5】直接切换到五笔输入法状态。

设置方法：

1. 在输入法标签（　）上单击鼠标右键，选择“设置”。

2. 弹出“文本服务和输入语言”对话框，选择“高级键设置”选项卡，单击“切换到中文—王码五笔型输入法 86 版”，如图 3—1—2 所示。

3. 单击“更改按键顺序”按钮，弹出如图 3—1—3 所示对话框。

4. 勾选“启用按键顺序”复选框，在左边下拉列表中选择“左【Alt】＋【Shift】”，在右边下拉列表中选择“5”，单击“确定”按钮。这时按下【Alt】＋【Shift】＋【5】即可直接切换到五笔输入法状态。

二、微软等拼音输入法中英文切换方法

对于目前比较流行的拼音输入法，如微软、百度、搜狗等，其中英文切换方法为：

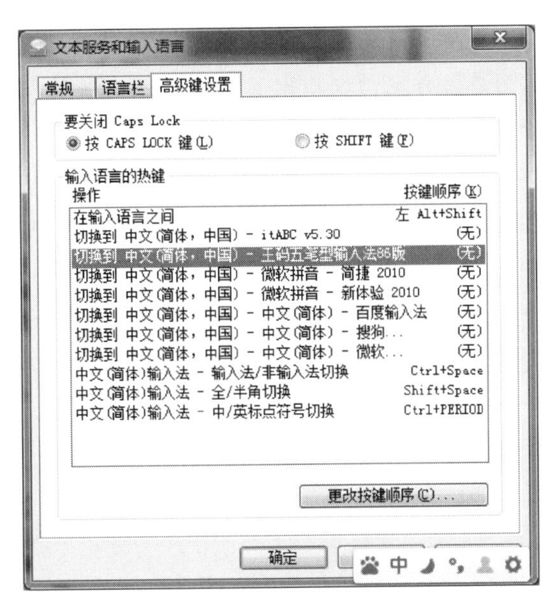

图 3—1—2 “文本服务和输入语言”对话框

更改按键顺序
切换到 中文(简体，中国) - 王码五笔型输入法86版
启用按键顺序(E)
左 Alt + Shift
键(K): 0
Ctrl
Ctrl + Shift
左 Alt + Shift
确定
取消

图 3—1—3 “按键选择”对话框

按下【Shift】键。

过关斩将

要求：以下测试内容，以 10 分钟为一个测试单元，必须达到 30 字/分钟（全部输完），才能进入下一环节的测试，全部测试完成后可以进入活动 2。

第一关：中英数录入练习（一段中文，一段英文，夹杂个别全角数字。）

新华网北京 8 月 28 日电 8 月 29 日人民日报评论员文章：在创新中赢得主动权——六论学习贯彻习近平总书记 8·19 重要讲话精神

“宣传思想工作创新，重点要抓好理念创新、手段创新、基层工作创新”，在全国宣传思想工作会议上的重要讲话中，习近平总书记深刻阐释了改革创新的重要性，系统提出了宣传思想工作重点要抓好的“三个创新”，为打好宣传思想工作主动仗明确了主攻方向，为赢得意识形态工作主动权指出了实现路径。在长期实践中，我们党的宣传思想工作积累了十分丰富的经验。

The shoemaker, however, could not endure a joke; he pulled a face as if he had drunk vinegar, and made a gesture as if he were about to seize the tailor by the throat. But the little fellow began to laugh, reached him his bottle, and said, “No harm was meant, take a drink, and swallow your anger down.” The shoemaker took a very hearty drink, and the storm on his face began to clear away. He gave the bottle back to the tailor, and said, “I spoke civilly to you; one speaks well after much drinking, but not after much thirst.”

第二关：中英数录入强化练习（一句中文，一句英文，夹杂个别全角数字，中文中夹杂英文符号如＄，＼，／，＠等，英文中夹杂中文符号如￥等。）

回望历史，找到这条正确道路，极为艰辛＼来之不易，它是在改革开放30多年的伟大实践中走出来的，是在中华人民共和国成立60多年的持续探索中走出来的，是在对近代以来170多年中华民族发展历程的深刻总结中走出来的，是在对中华民族5000多年悠久文明的传承中走出来的。这样的深厚历史渊源和广泛现实基础，使中国道路展现出旺盛的生命力，极大地增强了13亿人民的民族自信心和自豪感。站在过去与未来的梦想交汇点上，亿万人民的理论自信/道路自信/制度自信更加坚定。

For the April-to-June quarter, it grew at a rate of 4.4％, compared with the same period in the previous year. It was a weaker performance than most economists had been expecting and was a slowdown from the first three months of the year, when growth was 4.8％. A contraction in mining and manufacturing activity was behind the slowdown. Friday's figures show the economy is now expanding at the slowest rate since 2009. We do not wish to sound alarmist, but concern on the economy can hardly be overstated.

第三关：中英数录入综合练习（以中文为主，夹杂全角/半角的英文、数字等。）

4日早晨，吉林省长春市（Changchun City, Jilin Province）西环城路与隆化路交会处，一辆银灰色丰田RAV4轿车被盗。一同被“盗”的，还有车上一名仅两个月大的婴儿小皓博。小皓博的命运牵动着无数人的心，长春近万名警力和市民加入寻找行列。功夫不负有心人，5日早上，警方在四平地区发现被盗车辆，但未发现被“盗”婴儿。5日晚，吉林警方证实，犯罪嫌疑人于当天下午自首，承认掐死小皓博后埋于雪中。

孩子丢失，一座城寻一个孩子4日早7时20分左右，长春市民许先生仅两个月大的孩子随车被盗。许先生随即报警并向媒体求助，一场全城寻人的爱心接力赛上演。截至当晚 6 时许，长春市共出动警力 8 000 余人次，在全市范围内开展地毯式排查。

活动 2　无所不知——特殊符号录入

经过一个星期的刻苦训练，小文和小璐对于状态切换已经是驾轻就熟，现在无论是师傅随手给出的网络新闻、财经、体育文章，还是他们自己感兴趣的笑话、故事和小说，都可谓信手打来，再也不会“老虎吃天，无处下爪”了。

这天，师傅手里拿着两张纸走了进来：“来来来，徒儿，又有新任务了，你们把纸上的内容录入电脑吧！”

小文和小璐面带微笑一人接过一张纸，正待输入，细心的小璐发现了问题：“啊?! 这么多乱七八糟的符号，这该如何输入呀?”

小文凑过来一看：“是呀！太出格了！”

师傅看着俩人抓耳挠腮的样子，笑着摇了摇头：“你们俩啊，总是心急吃不了热豆腐，听为师给你们慢慢道来。”

摩拳擦掌

你们看到的这篇文章，里面涉及一些键盘上没有的符号，如图 3—2—1 圈中所示，这些符号就叫做特殊符号。

符号是人们共同约定用来指称一定对象的标志物，它可以包括以任何形式通过感觉来显示意义的全部现象。在这些现象中某种可以感觉的东西就是对象及其意义的体现者。例如：π,℃,№,▽,▲,◎,※,Ω

其中，大球形颗粒又称 Dane 颗粒，是 1970 年 Dane 首先用电镜在乙肝病人血清中发现的。Dane 颗粒是有感染性的完整 HBV 颗粒，呈球形，具有双层衣壳。外衣壳由来自宿主的脂质双层和包膜蛋白组成，有大约 400 个 HBV 表面抗原（HBsAg）即蛋白镶嵌于脂质双层中。用离子去垢剂如 NP-40 处理病毒颗粒，去除病毒外衣壳后，暴露出内层核心。核心的表面为病毒的内衣壳，内衣壳蛋白为 HBV 核心抗原（HBcAg）。HBcAg 经酶或去垢剂作用后可暴露出 e 抗原(HBeAg)。

图 3—2—1　含特殊符号的文章

关于特殊符号的录入，不同输入法有不同的录入方法。

师傅领进门

一、五笔86版输入法中特殊符号的输入

1. 在五笔状态条（五笔型）最右边的软键盘标志上单击鼠标右键。

2. 弹出下图3—2—2所示菜单。

3. 单击“特殊符号”，则在屏幕右下角出现如图3—2—2所示的特殊符号软键盘，输入相应字母即可得到特殊符号，比如输入【W】键可得到“№”；输入【F】可得到“※”等，如图3—3—3所示。

PC键盘	标点符号
希腊字母	数字序号
俄文字母	数学符号
注音符号	单位符号
拼　音	制表符
日文平假名	✔ 特殊符号
日文片假名	

图3—2—2　五笔86版特殊符号分类

图3—2—3　五笔86版特殊符号软键盘

4. 同样方法可得到拼音字母、数学符号、希腊字母、俄文字母、日文平假名、片假名等。

二、智能ABC输入法中特殊符号的输入

智能ABC输入法中可使用【V】键与数字1～9的组合完成特殊符号的输入，查找需要的特殊符号可通过【+】【-】键来翻页，找到之后输入特殊符号前面的数码即可。具体键组合与功能见表3—2—1。

表3—2—1　智能ABC输入法组合键功能表

键组合	功能
【V】+【1】	常用的特殊符号
【V】+【2】	数字
【V】+【3】	全半角数字、字母

续表

键组合	功能
【V】+【4】	日文平假名
【V】+【5】	日文片假名
【V】+【6】	希腊字母
【V】+【7】	俄文字母
【V】+【8】	汉语拼音字母
【V】+【9】	制表符

三、微软拼音输入法中特殊符号的输入

1. 在微软拼音状态条上单击“功能菜单”按钮，弹出菜单如图 3—2—4 所示。

2. 将鼠标移动到“软键盘”菜单，弹出菜单如图 3—2—5 所示。

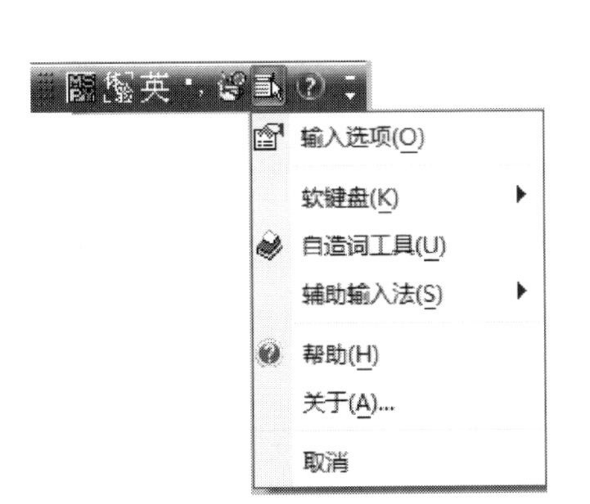

图 3—2—4 微软拼音功能菜单

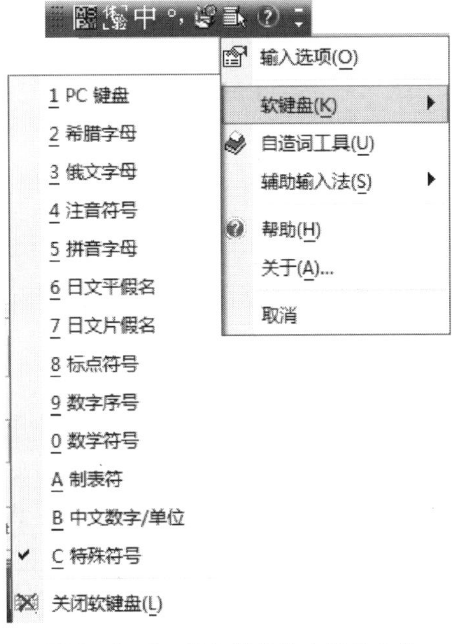

图 3—2—5 微软拼音特殊符号分类

3. 选择数字或字母来确定特殊符号的类型，比如输入【C】，则在屏幕右下角弹出特殊符号软键盘如图 3—2—6 所示，输入相应字母即可得到特殊符号。

图 3—2—6　微软拼音软键盘

四、搜狗拼音输入法中特殊符号的输入

1. 在搜狗拼音状态条（中 简 系）上单击鼠标右键，弹出菜单如图 3—2—7 所示。

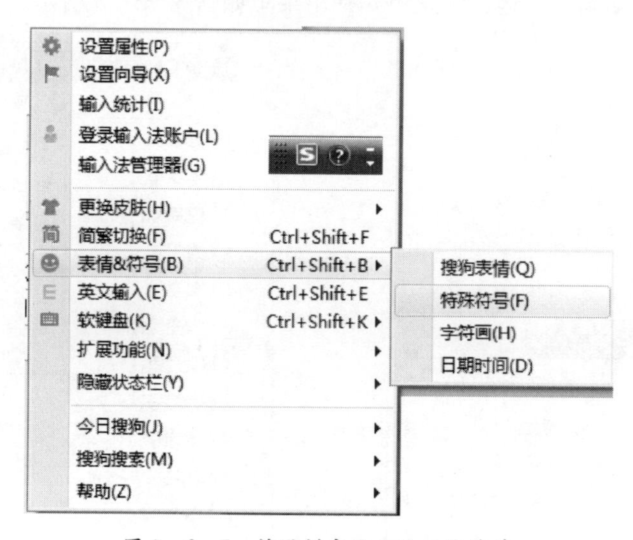

图 3—2—7　搜狗拼音输入法功能菜单

2. 单击“表情 & 符号”中的“特殊符号”，弹出如图 3—2—8 所示的对话框，先选择类别如“特殊符号”，然后单击自己需要的特殊符号即可录入。

五、百度拼音输入法中特殊符号的输入

1. 在百度拼音状态条（中）单击鼠标右键，弹出菜单如图 3—2—9 所示。

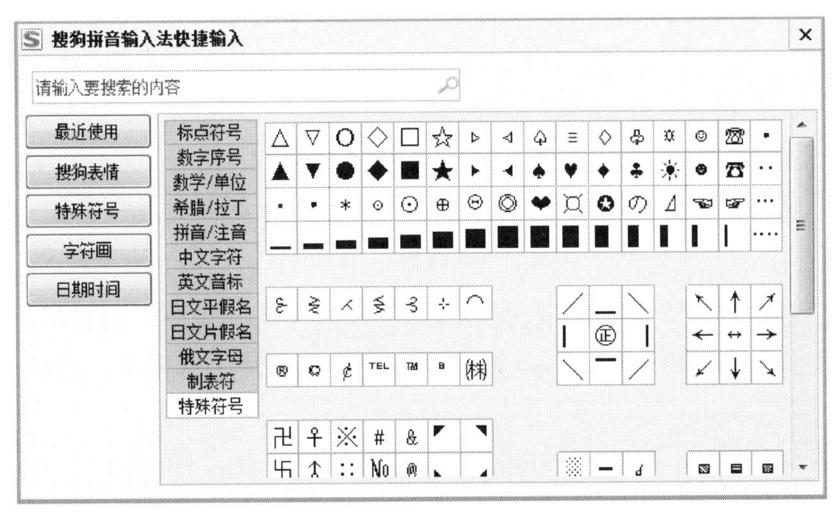

图 3—2—8 搜狗拼音输入法特殊符号分类及部分特殊符号

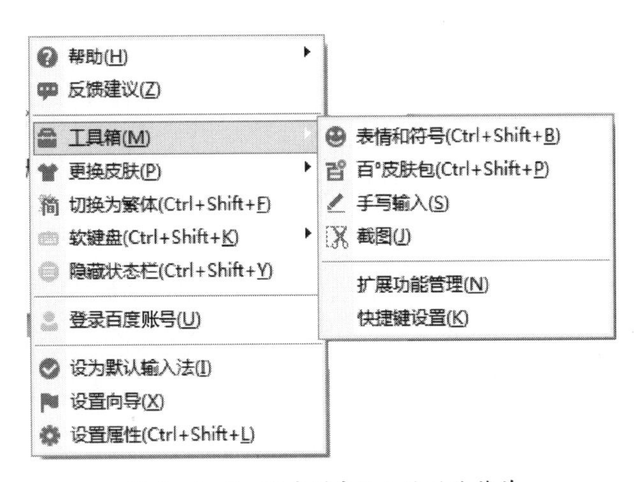

图 3—2—9 百度拼音输入法功能菜单

2. 单击“工具箱”中的“表情与符号”（或按下键盘上的【Ctrl】+【Shift】+【B】组合键），弹出如图 3—2—10 所示的对话框，先选择类别如“数字/单位”，然后单击自己需要的特殊符号即可录入。

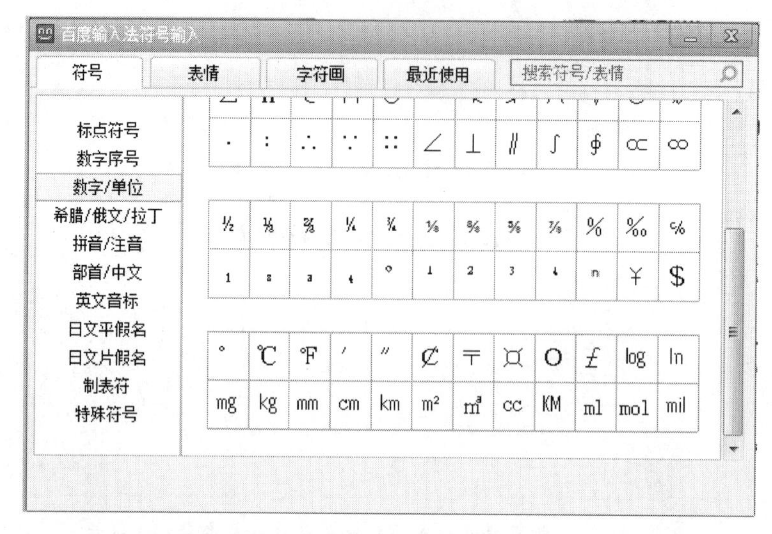

图 3—2—10 百度拼音输入法特殊符号分类及部分特殊符号

修行靠个人

要求：反复练习以下每部分的输入内容，直到该部分的目标达成以后，才能进入下一部分的输入练习！

一、使用五笔字型输入法进行特殊符号的录入

1. 输入以下特殊符号（基础训练：全部为特殊符号，一组切换一种类别。）

目标：录入速度达到 30 字符/分钟，即在 4 分钟内完成录入。

№↑@§ ￡℃¤‰ 〖〗『』 ≈≠∑∈ ∵∴//⊙∽ ㈤㈨ⅣⅩ

ニュシヒ みにゃち ǐāǔò ㄅㄐㄕㄡ пцги φαρπ

「‖」｛ ⅡⅥ㈦㈠ ÷∮∠√ ￠兆捌○ ┣┿┿┫ ☆▲◆§

τμζθ лбжя ㄋㄎㄝㄈ üúōè ゃふせし ムフホサ

2. 输入以下特殊符号（强化训练：全部为特殊符号，随机切换类别。）

目标：录入速度达到 30 字符/分钟，即在 4 分钟内完成录入。

サに∮」 ￡☆яθ 〖∑㈨ǔ ≈〗ふ兆 ∵Ⅹちㄎ ㈤』フⅣ

ニ@ヒǐ み↑ㄐ⊙ シāㄕ『 цㄅゃò пㄡ∴α φг▲┿

ュи‖｛ Ⅱ∽Ⅵ㈦ ÷㈠┿∠ ≠è捌○ ┣ρ§┫ ℃ホπ◆

τж‰ζ лбμ¤ ㄋ//úㄝ ㄈゃō￠ ∈üせし ム「√№

二、使用智能 ABC 进行特殊符号的录入

1. 输入以下特殊符号（基础训练：全部为特殊符号，一组切换一种类别。）

目标：录入速度达到30字符/分钟，即在4分钟内完成录入。

№↑÷～　(8)㈠ⅠⅪ　＊@￥C　ぁぎこぞ　イクチネ　ΓΠβθ

ЖЧФЯ　āòùǹ　━┆┐┣　…〖〗∧　︷︸ˊ　ΩΦγε

ˇ〔〕±　11.(4)⑩㈣　＂AMˆ　えぐちぬ　ォクスナ　ΔΠΣμ

ДШЪз　ěǒùā　┃┅┐┛　＊/@E　々‖『÷　18.㈠ⅡⅪ

2. 输入以下特殊符号（强化训练：全部为特殊符号，随机切换类别。）

目标：录入速度达到30字符/分钟，即在4分钟内完成录入。

№11.@±　(8)〗MΠ　＊チ÷ぁ　ぎβ￥ε　イC…ǹ　Γāこ┐

Ж～㈣∧　Πòθネ　━ù〖＂　┣クΦ㈠　︷μクγ　ΩぞΦ

ˇЧā〔　↑(4)Я⑩　A┆Ⅺˆ　Ⅱ┛ぐ‖　ォ︸┃18.　㈠Ⅰˆ Σ

Д〕ǒз　┅@ěЪ　Шスぬ┐　＊ù/E　々『ち÷　ナΔえⅪ

三、使用微软、百度、搜狗中的任一种进行特殊符号的录入

1. 输入以下特殊符号（基础训练：全部为特殊符号，一组切换一种类别。）

目标：录入速度达到30字符/分钟，即在4分钟内完成录入。

@•﹏¿　㈠⒁18.ⅵ÷≈≌¼℃¤￡㏖ξ ε Φæáéīù

ㄎㄘㄒㄌ　ㄥㄋㄔㄩəːiəʒtʃひむぜぐグヒテゼ┏━┓

☆☜▼♣◎¤⊙⊕㊥㊧由№♫¶♪『』︷︸⒀3.❺ⅶ

±≠∑∮π σ и ㎡óńaǖ〤 刂氵 廴ʒəːɔː ʌ√♀☆の

2. 输入以下特殊符号（强化训练：全部为特殊符号，随机切换类别。）

目标：录入速度达到30字符/分钟，即在4分钟内完成录入。

•ㄘㄔ㈠　¿18.əː≈¤ⅵΦ¼∮ù÷éξ≌ εæá㏖━￡

ㄎ@☜⊙﹏ヒ¤iə⒁ㄋむʒひ tʃ┐グㄥぜ♀︷┏√3.ぐ

☆≠⊕ㄌ　◎ㄩ♣ㄒ㊥a♪由♫』№ǖ『♪テ︸⒀ńの 刂

▼и℃∑σ±ʌīó㊧❺¶〤ɔːⅶ氵 ʒπ廴əː㎡ゼ━☆

灵丹妙药

一、文章中特殊符号录入技巧

在一篇文章中，特殊符号所占比例很小，如果遇到一个特殊符号就进行切换输入会很浪费时间，建议在整个文章录入基本完成之后再进行特殊符号的录入。录入时把所有的特殊符号都一次性录入，然后再分别移动到其应处的位置。

二、在搜狗和百度拼音输入方法中也可以使用【V】键和1～9数字的组合来完成特殊符号的录入。

过关斩将

要求：以下测试内容，使用五笔字型输入法来输入特殊符号，必须达到 30 字/分钟（每一关有完成时间提示），才能进入下一环节的测试，全部测试完成后可以进入活动 3。

第一关：特殊符号录入练习（全部为特殊符号，在 3 分钟以内完成可过关。）

サに∮」£☆яθ〖∑⑼ǔ≈〗ふ兆∵Ⅹちㄎ㈤』フⅣニ@ヒǐみ↑ㄐ⊙シāㄕ『цㄅやòпㄡ∴αφг▲┼ɹи‖｛∽Ⅵ㈦÷㈠┼∠≠è捌○├ρ§┤℃ホπ◆τж‰ζлбμ¤ㄋ∥úせㄈやō￠∈üせしム「√№

第二关：特殊符号强化练习（一行中文，一行特殊符号，在 7 分钟以内完成可过关。）

学生进校的第一学期我们已经将五笔录入知识全部教授，学生也基本能利用
üせαサ∠≠ホㄕ『』бφ√№フⅣ∽Ⅵ㈤ニ@☆я÷㈠℃ㄅ≈ɹúせㄎやㄡ
五笔录入的方法进行中文文章的录入，当然学生录入的速度取决于其是否完
↑ㄐ⊙è○π◆τ〗ふㄈжиやム「┼∵Ⅹā‖ō‰ζлг▲」£θ〖∑⑼ǔ¤
全按照教师及班主任的方法进行。第二学期学生进入强化训练阶段，在该阶
ㄋ∥￠∈§┤し√τ｛⑺シц№∴ρò∠пヒǐみ├μ┼∮≠㈠∵≈φ⊙φ@
段学生会出现高原平台现象。

第三关：特殊符号录入综合练习（以中文为主，夹杂特殊符号，在 11 分钟以内完成可过关。）

符号是人们共同约定用来指称一定对象的标志物，它可以包括以任何形式通过感觉来显示意义的全部现象。在这些现象中某种可以感觉的东西就是对象及其意义的体现者。例如：π,℃，№，▽，▲，※，Ω 等。

其中，大球形●颗粒又称 Dane 颗粒，是 1970 年 Dane 首先用电镜在乙肝病人血清中发现的。Dane 颗粒是有感染性的完整 HBV 颗粒，呈球形●，具有双层衣壳◎。外衣壳○由来自宿主的脂质双层和包膜蛋白组成，有大约 400 个 HBV 表面抗原【HBsAg】即蛋白镶嵌于脂质双层中。用离子去垢剂如 NP-40 处理病毒颗粒，去除病毒外衣壳后，暴露出内层核心。核心的表面为病毒的内衣壳，内衣壳蛋白为 HBV 核心抗原【HBcAg】。HBcAg 经酶或去垢剂作用后可暴露出 e 抗原〖HBeAg〗。

活动 3　大功告成——文章练习

学期快要结束了，经过前段时间的磨炼，小文和小璐的心智成熟了很多，也沉稳了很多，他们一如既往地努力学习、训练着，达到了一个又一个新目标，期待着师傅给予他们的一个又一个挑战，他们不再自得、不再骄傲，师傅的言传身教让他们明白“山外有山，人外有人”，只有持续不断地、耐心地、科学地学习和训练才能到达理想的高度。

今天，师傅又会带给他们怎样的“惊喜”呢？

师傅：“今天考考你们的观察力，平时你们看到的文章中除了汉字还可能有哪些字符呢？”

小文和小璐争先恐后：“还有标点符号、英文字母、数字、特殊符号……”

师傅：“嗯，不错！观察得比较到位，今天我们就来进行最后一个训练——文章录入。”

摩拳擦掌

从开始学习文字录入到现在，你们已经学会了英文指法、数字录入、中文录入方法、状态切换、特殊符号录入等，那么在实际的文章录入当中你们可能还会遇到各种各样的问题，我们就根据前面学过的知识，兵来将挡、水来土掩。

师傅领进门

一、英文指法

按照项目一“英文录入”的指法要求，在保证准确率的前提下，坚持“盲打”。

二、文章中个别数字的输入

要求使用主键盘区上面的一排数字键进行录入，各手指管辖数字键见表 3—3—1。

表 3—3—1　　录入个别数字时手指分工表

手指	数字键
左手小指	1
左手无名指	2
左手中指	3
左手食指	4、5

续表

手指	数字键
右手食指	6、7
右手中指	8
右手无名指	9
右手小指	0

三、中文输入注意事项

● 注意指法，坚持盲打。

● 牢记特殊中文标点符号的输入方法，如省略号（……）、破折号（——）、分隔符（·）、人民币（￥）等。

● 注意录入过程中的状态切换，如中英文输入法的切换（【Ctrl】＋【Space】），中英文标点符号的切换（【Ctrl】＋【.】），全角/半角的切换（【Shift】＋【Space】，只有在中文文章中出现个别的全角英文和数字时会用到）等。

● 注意录入过程中个别大写英文字母的录入，可直接用左手小指按下【Caps Lock】键进行切换。

● 能用词组录入的一定要使用词组功能，能用多字词录入的绝不要用少字词，哪怕一开始慢一点，等时间长了，积累的词组会越来越多，打字速度会越来越快。

● 对于文章中出现的特殊符号，可先放过，等文章基本录完之后再进行插入。

修行靠个人

要求：反复练习录入以下文章内容，设置练习时间为10分钟，要求每录完一遍都要记录练习的现状，进行反思，找出训练过程中存在的问题，要求达到初级目标，争取达到高级目标。初级：30字/分钟；中级：40字/分钟；高级：60字/分钟。

文章内容	累计字数
有这样一个真实的故事：一行中国人在欧洲一家餐厅用餐后，	27
由于桌上剩了1/3的饭菜被罚款£50。工作人员解释道，虽然吃饭	58
的钱是你们的，但资源是社会的，谁都没有理由浪费。在全社会掀	87
起反对浪费的热潮中，这则故事发人深省。常听到有人振振有词：	116
花钱消费是个人自由，浪不浪费他人无权干涉。殊不知，浪费行为	145
的“溢出效应”侵蚀了社会资源，浪费就不仅是个人领域的事情，	174
更关乎公共利益；厉行节约也不仅是一种个人私德，更是一种社会	203
公德，从这个角度看，厉行节俭、人人有责。	223

续表

文章内容	累计字数
今天，中国跻身于世界第2大经济体，生活富了、腰包鼓了，	250
不少人认为浪费一点也无妨。的确，随着人们的消费习惯从“生存	280
型”转向“发展型”，适当提高消费水平可以理解，但这绝不能成为	310
丢掉勤俭这个传家宝的理由。且不论农村仍有一亿多扶贫对象，也	339
不说人均GDP在世界排名仍然靠后，任何时候富裕都不是浪费的通行	370
证，我们又怎能未富先奢？	380
也有一种浪费行为，源于扭曲的消费观。或是“面子消费”，	407
认为大操大办、宁剩毋缺、宁多毋少面子上才挂得住；或是“炫耀	436
消费”，不选对的、专炒贵的，以此炫耀财富、夸饰身份、显示地位。	467
任由扭曲的消费观滋生蔓延，就会产生“绑架”现象，讲面子、好	496
排场、相互攀比，大家这样、我也这样，必然助长奢靡之风。然而，	526
个人虚荣满足了，社会资源却浪费了，这笔账怎么算？	550
经济学家说过，奢侈是公众的大敌，节俭是社会的恩人。中国	577
是一个人口大国，如果人人躬行节俭，那将是多么巨大的财富？	605

填表说明：在表3—3—2中记录练习的速度、回退率、码长以及错字，找到回退率高、码长较长以及错字多的原因，并对错字进行订正。

表3—3—2　　文章练习记录表

项目	练习现状记录	存在问题/订正
速度		
回退率		
码长		
错字		

灵丹妙药

一、怎样提速比较科学

到了最后这个阶段，学生已经掌握了文字录入的基本方法，进入提速训练阶段。这时，由于文章的难度有所提升（中文、英文、数字、全角/半角字符均可能包含），大部分学生的差错率会比较高，使用回退键的次数明显增多，这时切忌盲目追求速度，应把准确率放在第一位，坚持盲打，在提高准确率的前提下逐步提速。这个阶段的学生应在老师指导下制定适合自己的阶段性目标，以稳为主。

二、在文章练习中如何找到自己的“短板”

录完一篇文章后，如果成绩不理想，不要急着再练，而是结合文章练习记录表，找到自己的“短板”，有针对性地加以解决。

- 回退率高：说明击键的准确性不够，通过录屏软件把打字视频拍下来，然后回放，把容易使用回退键的字摘录出来，反复练习，直到达到零回退。
- 码长较长：简码和词组的应用有问题，要逐字过关，坚持能打词组的一定要使用词组，能使用二级、三级简码的绝不使用四码。
- 错字较多：说明拆字有问题，准备一个错字本，把错字及其拆分方法写出来，并录入到打字机反复练习，直到不再打错为止。
- 熟的文章快、生的文章慢：需要找打字的手感，一篇文章反复打，打到无法提速了，再去打第二篇，这样慢慢找到手感。
- 回退率低、错字很少，码长适当，但速度提不上去：建议每天抽 30 分钟左右的时间练习录入英文，提高击键速度。

三、辅助练习软件

1. 爱不释手 1.82 版。该软件的特点：测试结果详尽，除了普通的速度、准确率外，利用软件提供的指法检测功能可以得到您对每个字符的平均反应时间及易错字符等信息，据此进行更有针对性的练习。

2. QQ 群跟打练习。目前网上比较知名的打字群有：精五门（47733203）、舞指·爱（72129639）。在群里会有人发文章，大家跟着打即可。打完一段文章后会有如下显示，如：第 10285 段、速度 103.37、回改 2、错字 0、击键 4.41、码长 2.56、单字 60%等。在这样的一个 QQ 群里跟打，会提高学生五笔练习的积极性和趣味性。

过关斩将

要求：下面的三篇文章由易到难，每篇文章按照 10 分钟的时限进行练习和测试，每次完成后记录下累计录入的总字数，通过反复练习直至达到既定目标。达到初级目标（30 字/分钟）之后再尝试去达到中级目标（40 字/分钟），最后通过自己的努力达到高级目标（60 字/分钟）。一篇练完后再练习下一篇。注意：段落随机。

第一篇：政论文章

续表

文章内容	累计字数
共筑中国梦 Chinese Dream，需要经济社会的不断发展，需要民	33
生的持续改善，这是复兴之本、梦想之基。	52
随时随刻倾听人民呼声、回应人民期待，使发展成果更多更公	79
平惠及全体人民，朝着共同富裕方向稳步前进。习近平主席的重要	108
讲话，阐明了中国梦 Chinese Dream 的丰富内涵，凸显出民生工作和	143
社会管理在实现中国梦过程中极其重要的作用。	164
从“贫穷不是社会主义”到“共同富裕”，从“发展是硬道理”	192
到“全面建成小康社会”，60 多年来，发展经济，改善民生，始终	222
是党和政府最重要的工作。中国改革发展的历程，正是在倾听人民	251
呼声、回应人民期待中不断深化的，也必将沿着这个方向继续推进，	281
不断实现好、维护好、发展好最广大人民的根本利益，让老百姓得	310
到更多好处。	315
中国梦 Chinese Dream，是对公平正义的向往。实现公平正义，	348
既需要处理好教育、就业、养老、医疗、收入分配等方面的问题，	377
也需要权利保障更加充分、人人得享共同发展。只有这些得到妥善	406
解决，社会才能安定有序，国家才能长治久安。正如习近平主席所	435
强调的，“保证人民平等参与、平等发展权利，维护社会公平正义”，	466
这是实现中国梦的重要内容。	479
中国梦 Chinese Dream，是对改善生活的渴望。这些年来，无论	512
是城乡居民收入持续增长、社会保障网初步建立，还是医疗、教育	541
等方面不断向前的改革，人民群众的幸福指数不断提高。农业税免	570
除了，义务教育免费了，职工工资增加了，城市低保标准提高了，	599
农民有了基本医疗保障。	610

第二篇：小说散文/科技论文

文章内容	累计字数
纳米（nanometer）是一个微小的长度单位，1 纳米等于十亿	29
分之一米。1 根头发丝有 7 万到 8 万纳米。纳米技术这个词汇出	62
现在 1974 年。纳米科学、纳米技术是在 0.10 到 100 纳米尺度的空	95
间内研究电子、原子和分子的运动规律及特性。纳米材料是纳米技术	124
的重要组成部分，也是国际上竞争的热点和难点。碳纳米管自从	153
1991 年被发现以来，就一直被誉为未来的材料。碳纳米管在强度上	183
大约比钢强 100 倍，其传热性能优于所有已知的其他材料。碳纳米	213
管具有良好的导电性，在常温下导电时，几乎不产生电阻。纳米陶	242
瓷材料在 1600 摄氏度高温下能像橡皮泥那样柔软，在室温下也能自	272

续表

文章内容	累计字数
由弯曲。从1998年世界上第一支纳米晶体管制成，到1999年100	302
纳米芯片问世，使20世纪最后10年世界上出现的“纳米热”进一	334
步升温。	345
我国在纳米技术领域占有一席之地，处于国际先进行列。已成	372
功制备出包括金属、合金、氧氢化物、氢化物、碳化物、离子晶体	401
和半导体等多种纳米材料，合成出多种同轴纳米电缆，掌握了制备	430
纯净碳纳米管技术，能大批量制备长度为2至3毫米的超长纳米管。	460
合成的最细的碳纳米管的直径只有0.33纳米，这不但打破了我国科	491
学家自己不久前创造的直径只为0.5纳米的世界纪录，而且突破了	521
日本科学家1992年所提出的0.4纳米的理论极限值，《稻草变黄金	553
——从四氯化碳制成金刚石》的文章对其进行了高度评价。最近又	582
研制成功新型纳米材料——超双疏性界面材料。	603

第三篇：模拟“星光计划”文字录入比赛文章

文章内容	累计字数
本文对ICU重症抢救性气管插管患者采用选择性气管插管前镇	27
静，即插管有抵抗患者用咪唑安定（或）丙泊酚镇静，插管无抵抗	56
患者不用镇静，其一次性气管插管成功率为93.02%，无呼吸心跳骤	87
停及恶性心律失常发生。发生反流误吸、呼吸暂停、支气管痉挛、	116
低血压病例数略少于宋莉等［1］的报道。Knaus等［5］于1985年	152
提出了APACHE II评分系统并在国际上广泛应用，国内报道A-	182
PACHE II评分系统可应用于评估病情程度和预后的依据［6-7］。	217
妃嫔媵嫱，王子皇孙，辞楼下殿，辇来于秦。朝歌夜弦，为秦	244
宫人。明星荧荧，开妆镜也；绿云扰扰，梳晓鬟也；渭流涨腻，弃	273
脂水也；烟斜雾横，焚椒兰也。雷霆乍惊，宫车过也；辘辘远听，	302
杳不知其所之也。一肌一容，尽态极妍，缦立远视，而望幸焉，有	331
不得见者，三十六年！燕赵之收藏，韩魏之经营，齐楚之精英，几	360
世几年，摽掠其人，倚叠如山；一旦不能有，输来其间；鼎铛玉石，	390
金块珠砾，弃掷逦迤，秦人视之，亦不甚惜。	410
计一次气管插管成功（判断标准：气管导管进入距门齿女20 cm,	441
男22 cm以上视为插管一次）的例数；计插管时抵抗插管需用	470
镇静例数，插管时患者心跳骤停、反流误吸、呼吸暂停、支气管痉挛	499
（判断标准：插管前后听肺部干啰音从无到有或者干啰音明显增	528
加）、低血压［判断标准：正常血压者收缩血压下降至小于90 mmHg	562
（1 mmHg=0.133 kPa）、高血压者收缩血压下降大于30 mmHg）］。	605

藏经阁

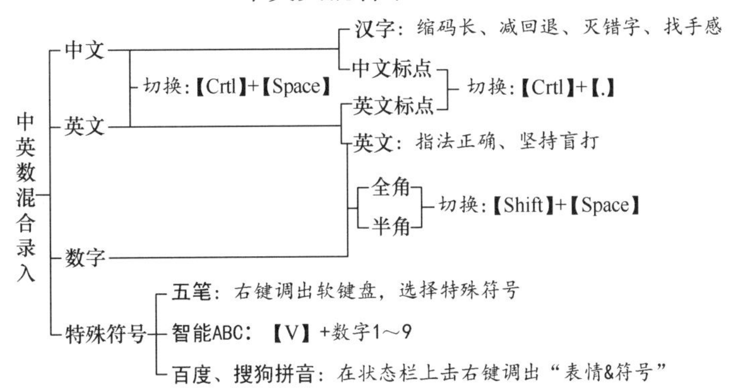

打擂台

以下混合录入文稿需要录入,记录完成的时间,并参照下表给出自我评价。中级水平的录入速度 60 字符/分钟是每个人通过短期努力都可以达到的。

混合录入自我评价表

级别	完成时间(分钟)	录入速度(字/分钟)	称号	文字录入四级鉴定标准
入门	32	20	打字新手	—
初级	21	30	打字老手	—
中、低级	16	40	打字熟手	合格
中级	10	60	打字高手	良好
中、高级	8	80	打字健将	优秀
高级	5	120	打字达人	—

测试文章

文章内容	累计字数
东方网 10 月 17 日消息:记者(reporter)昨天获悉,今年	31
5 月以来,上海公安机关共侦破发票(invoice)类犯罪 800 余起,	65
刑事拘留 700 余人,缴获各类假发票 140 余万份,捣毁制售假发票、	97
开票窝点 19 处,直接挽回国家税款损失 6 200 余万元。	123
今年 8 月,徐汇公安在工作中发现:上海某物流有限公司负责人	151

续表

文章内容	累计字数
顾某有利用国家“营改增”政策，对外虚开增值税专用发票的重大	180
嫌疑。随着调查的一步步深入，一个价税逾亿的案件（case）逐步	211
浮出水面。	216
今年1月至7月，犯罪嫌疑人顾某伙同他人控制多家公司	241
（company）。这些公司没有发生真实业务，却以收取票面金额5%	274
“开票费”的方式，为7个省市的200余家单位虚开增值税专用发	304
票700余份。为掩盖虚开发票的犯罪事实，顾某又通过支付票面金	334
额1.6%～1.7%“开票费”的方式，先后让3个省市的20余家企业	366
虚开增值税专用发票180余份。	382
据公安部门介绍，虚开发票案件中，一些中介公司扮演着重要	409
角色。这些中介机构在各区县经济开发园区内设立企业登记、企业	438
登记代理、企业投资管理、代理记账等类型的公司，实质上却从事	467
介绍虚开增值税专用发票等犯罪的中介业务（business）。	497
公安部门表示，增值税专用发票的开具需严格遵守票、货、款	524
一致原则，为他人虚开、接受他人虚开、介绍他人虚开、为自己虚	553
开增值税专用发票的行为均属违法。作为中华人民共和国公民，每	582
个人都有义务管好自己、监督他人，发现有倒买倒卖发票、非法代	611
开发票以及贩卖假发票的，要及时举报给公安机关，以帮助公安机	640
关及时破案。	646

ULUYI

附录一 成字字根编码表

（可用全码、简码输入）

汉字	全码	简码	汉字	全码	简码	汉字	全码	简码	汉字	全码	简码
戋	GGGT		廿	AGHg	AGH	竹	TTGh	TTG	巳	NNGN	
五	GGhg	GG	七	AGn	AG	手	RTgh	RT	己	NNGn	NNG
士	FGHG		上	Hhgg	H	斤	RTTh	RTT	尸	NNGT	
二	FGg	FG	止	HHhg	HH	乃	ETN		心	NYny	NY
干	FGGH		卜	HHY		用	ETnh	ET	羽	NNYg	NNY
十	FGH		曰	JHNG		豕	EGTy	EGT	耳	BGHg	BGH
寸	FGHY		早	JHnh	JH	八	WTY		了	Bnh	B
雨	FGHY		虫	JHNY		儿	QTn	QT	也	BNhn	BN
犬	DGTY		川	KTHH		夕	QTNY		孑	BNHG	
三	DGgg	DG	甲	LHNH		文	YYGY		刀	VNt	VN
古	DGHg	DGH	四	LHng	LH	方	YYgn	YY	九	VTn	VT
石	DGTG		皿	LHNg	LHN	广	YYGT		臼	VTHg	VTH
厂	DGT		车	LGnh	LG	辛	UYGH		巴	CNHn	CNH
丁	SGH		力	LTn	LT	六	UYgy	UY	马	CNng	CN
西	SGHG		由	MHng	MH	门	UYHn	UYH	弓	XNGn	XNG
戈	AGNT		贝	MHNY		小	IHty	IH	匕	XTN	
弋	AGNY		几	MTn	MT	米	OYty	OY	幺	XNNY	

ULUER

附录二 末笔识别码编码表

A：艾（AQU） 皑（RMNN） 岸（MDFJ） 凹（MMGD）
B：叭（KWY） 扒（RWY） 笆（TCB） 疤（UCV） 把（RCN）
坝（FMY） 泵（DIU） 柏（SRG） 败（MTY） 拌（RUFH）
剥（VIJH） 卑（RTFJ） 钡（QMY） 狈（QTMY） 叉（CYI）
备（TLF） 铂（QRG）
C：仓（WBB） 草（AJJ） 厕（DMJK） 叉（CYI） 岔（WVMJ）
忏（NTFH） 扯（RHG） 彻（TAVN） 尘（IFF） 程（TKGG）
驰（CBN） 尺（NYI） 斥（RYI） 愁（TONU） 仇（WVN）
丑（NFD） 臭（THDU） 床（YSI） 闯（UCD） 辞（TDUH）
触（QEJY） 囱（TLQI）
D：歹（GQI） 待（TFFY） 丹（MYD） 悼（NHJH） 等（TFFU）
笛（YMF） 狄（QTOY） 翟（NWYF） 刁（NGD） 叮（KSH）
冬（TUU） 斗（UFK） 抖（RUFH） 杜（SFG） 肚（EFG）
妒（VYNT） 兑（UKQB）
E：讹（YWXN） 厄（DBV） 尔（QIU） 洱（IBG） 铒（QBG）
F：伐（WAT） 乏（TPI） 钒（QMYY） 坊（FYN） 肪（EYN）
仿（WYN） 访（YYN） 飞（NUI） 吠（KDY） 奋（DLF）
忿（WVNU） 粪（OAWU） 封（FFFY） 孚（EBF） 拂（RXJH）
伏（WDY） 弗（XJK） 付（WFY） 父（WQU） 讣（YHY）
G：改（NTY） 甘（AFD） 杆（SFH） 竿（TFJ） 赶（FHFK）
秆（TFH） 冈（MQI） 皋（RDFJ） 杲（JSU） 告（TFKF）
恭（AWNU） 汞（AIU） 勾（QCI） 苟（AQKF） 辜（DUJ）
咕（KDG） 沽（IDG） 蛊（JLF） 故（DTY） 固（LDD）
刮（TDJH） 挂（RFFG） 圭（FFF）
H：旱（JFJ） 汗（IFH） 夯（DLB） 豪（YPEU） 亨（YBJ）
弘（XCY） 户（YNE） 幻（XNN） 皇（RGF） 惶（NRGG）
回（LKD） 茴（ALKF） 卉（FAJ） 昏（QAJF） 荤（APLJ）
霍（FWYF）
J：击（FMK） 讥（YMN） 伎（WFCY） 芰（AFCU） 剂（YJJH）
荠（AYJJ） 忌（NNU） 佳（WFFG） 贾（SMU） 钾（QLH）
笺（TGR） 肩（YNED） 奸（VFH） 茧（AJU） 贱（MGT）
见（MQB） 涧（IUJG） 饯（QNGT） 溅（IMGT） 疖（UBK）

秸（TFKG） 劫（FCLN） 戒（AAK） 诫（YAAK） 巾（MHK）
竞（UKQB） 今（WYNB） 筋（TELB） 仅（WCY） 京（YIU）
惊（NYIY） 井（FJK） 酒（ISGG） 巨（AND） 句（QKD）
眷（UDHF） 卷（UDBB） 抉（RNWY） 诀（YNWY） 钧（QQUG）
君（VTKD）

K：卡（HHU） 刊（FJH） 看（RHF） 尻（NVV） 抗（RYMN）
亢（YMB） 苦（ADF） 库（YLK） 框（SAGG） 矿（DYT）
旷（JYT） 亏（FNV）

L：垃（FUG） 兰（UFF） 泪（IHG） 厘（DJFD） 莅（AWUF）
笠（TUF） 栗（SSU） 溧（ISSY） 痨（UDNV） 粒（OUG）
利（TJH） 隶（VII） 连（LPK） 凉（UYIY） 晾（JYIY）
掠（RYIY） 疗（UBK） 吝（YKF） 漏（INFY） 芦（AYNR）
庐（YYNE） 虏（HALV） 仑（WXB）

M：玛（GCG） 吗（KCG） 码（DCG） 蚂（JCG） 麦（GTU）
忙（NYNN） 卯（QTBH） 冒（JHF） 枚（STY） 眉（NHD）
美（UGDU） 闷（UNI） 盂（BLF） 苗（ALF） 庙（YMD）
灭（GOI） 茗（AQKF） 闽（UJI） 牡（RTFG） 亩（YLF）
沐（ISY）

N：艿（AEB） 柰（SFIU） 尿（NII） 捏（RJFG） 涅（IJFG）
牛（RHK） 农（PEI） 弄（GAJ） 疟（UAGD）

O：呕（KAQY）

P：判（UDJH） 刨（QNJH） 匹（AQV） 票（SFIU） 迫（RPD）
粕（ORG） 扑（RHY） 朴（SHY） 钋（QHY）

Q：栖（SSG） 奇（DSKF） 企（WHF） 气（RNB） 乞（TNB）
泣（IUG） 讫（YTNN） 千（TFK） 扦（RTFH） 仟（WTFH）
浅（IGT） 巧（AGNN） 羌（UDNB） 茄（ALKF） 妾（UVF）
怯（NFCY） 青（GEF） 琼（GYIY） 丘（RGD） 酋（USGD）
蛆（JEGG） 去（FCU） 泉（RIU） 雀（IWYF）

R：冉（MFD） 壬（TFD） 仁（WFG） 刃（VYI） 戎（ADE）
茸（ABF） 冗（PMB） 汝（IVG）

S：杀（QSU） 晒（JSG） 腮（ELNY） 杉（SET） 钐（QET）
汕（IMH） 扇（YNND） 尚（IMKF） 勺（QYI） 舌（TDD）

	申（JHK）	声（FNR）	升（TAK）	圣（CFF）	什（WFH）
	矢（TDU）	屎（NOI）	仕（WFG）	市（YMHJ）	谁（YWYG）
	私（TCY）	宋（PSU）	诵（YCEH）	酥（SGTY）	粟（SOU）
	岁（MQU）				
T：	她（VBN）	坍（FMYG）	叹（KCY）	讨（YFY）	套（DDU）
	贴（MHKG）	汀（ISH）	廷（TFPD）	童（UJFF）	头（UDI）
	秃（TMB）	徒（TFHY）	吐（KFG）	推（RWYG）	驮（CDY）
W：	洼（IFFG）	万（DNV）	丸（VYI）	亡（YNV）	枉（SGG）
	旺（JGG）	妄（YNVF）	唯（KWYG）	未（FII）	位（WUG）
	蚊（JYY）	纹（XYY）	紊（YXIU）	问（UKD）	沃（ITDY）
	吾（GKF）	捂（RGKG）	毋（XDE）	午（TFJ）	忤（NTFH）
	迕（TFPK）	伍（WGG）	勿（QRE）	悟（NGKG）	
X：	昔（AJF）	硒（DSG）	矽（DQY）	汐（IQY）	虾（JGHY）
	匣（ALK）	闲（USI）	香（TJF）	湘（ISHG）	乡（XTE）
	翔（UDNG）	享（YBF）	泄（IANN）	屑（NIED）	芯（ANU）
	忻（NRH）	锌（QUH）	囟（TLQI）	杏（SKF）	刑（GAJH）
	兄（KQB）	汹（IQBH）	朽（SGNN）	玄（YXU）	穴（PWU）
	血（TLD）	驯（CKH）			
Y：	丫（UHK）	岩（MDF）	阎（UQVD）	厌（DDI）	唁（KYG）
	彦（UTER）	秧（TMDY）	羊（UDJ）	佯（WUDH）	仰（WQBH）
	杳（SJF）	舀（EVF）	耶（BBH）	曳（JXE）	沂（IRH）
	艺（ANB）	邑（KCB）	亦（YOU）	异（NAJ）	羿（NAJ）
	翌（NUF）	音（UJF）	尹（VTE）	页（DMU）	应（YID）
	拥（REH）	佣（WEH）	痈（UEK）	蛹（JCEH）	酉（SGD）
	尤（DNV）	疣（UDNV）	昱（JUF）	元（FQB）	圆（LKMI）
	云（FCU）	芸（AFCU）	孕（EBF）	誉（IWYF）	
Z：	闸（ULK）	札（SNN）	扎（RNN）	盏（GLF）	章（UJJ）
	丈（DYI）	仗（WDYY）	瘴（UUJK）	兆（IQV）	召（VKF）
	皂（RAB）	砧（DHKG）	正（GHD）	置（LFHF）	痔（UFFI）
	钟（QKHH）	仲（WKHH）	舟（TEI）	诌（YQVG）	肘（EFY）
	住（WYGG）	爪（RHYI）	庄（YFD）	壮（UFG）	状（UDY）
	坠（BWFF）	谆（YYBG）	啄（KEYY）	卓（HJJ）	孜（BTY）
	仔（WBG）	自（THD）	汁（IFH）	走（FHU）	足（KHU）

ULUSAN

附录三 常用千字

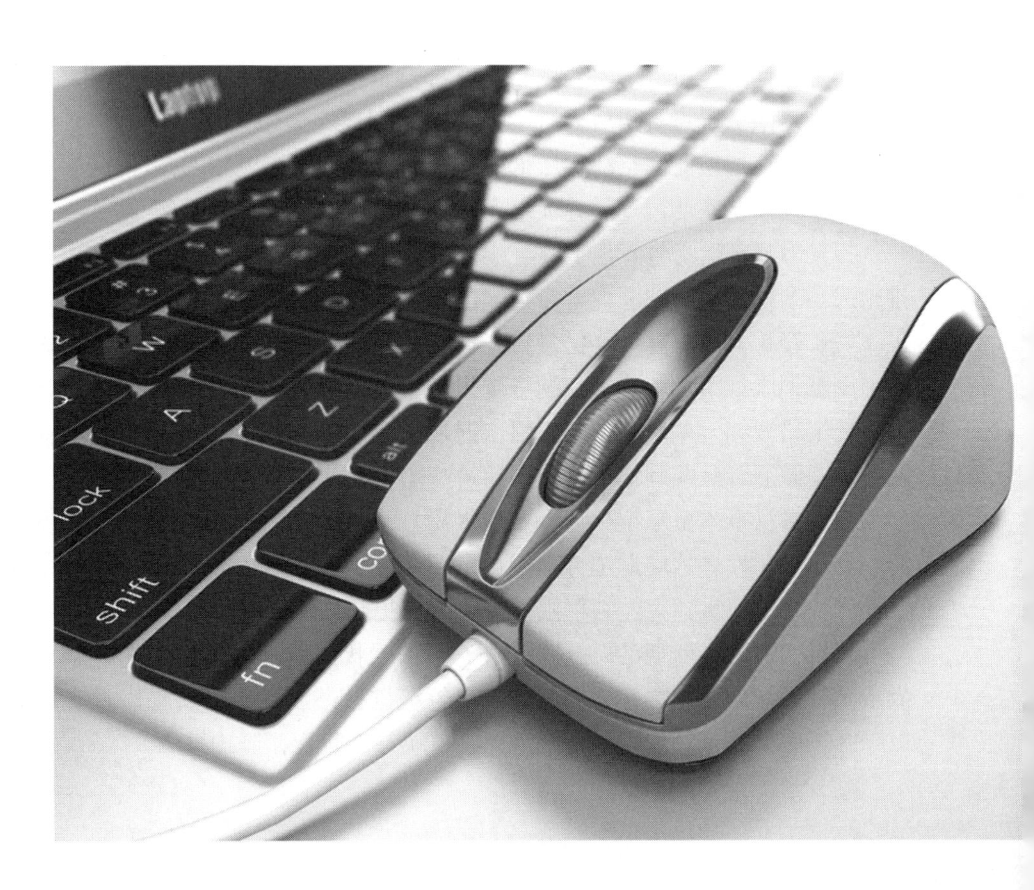

的一国在人了有中是年和大业不为发会工经上地市要个产这出行作生家以成到日民来我部对进多全建他公开们场展时理新方主企资实学报制政济用同于法高长现本月定化加动合品重关机分力自外者区能设后就等体下万元社过前面

农也得与说之员而务利电文事可种总该三各好金第司其从平代当天水省提商十管内小技位目起海所立已通入量子问度北保心还科委都术使明着次将增基名向门应里美由规今题记点计去强两些表系办教正条最达特革收二期并程厂如

道际及西口京华任调性导组东路活广意比投决交统党南安此领结营项情解议义山先车然价放世间因共院步物界集把持无但城相书村求治取原处府研质信四运县军件育局干队团又造形级标联专少费效据手施权江近深更认果格几看没

职服台式益想数单样只被亿老受优常销志战流很接乡头给至难观指创证织论别五协变风批见究支那查张精林每转划准做需传争税构具百或才积势举必型易视快李参回引镇首推思完消值该走装众责备州供包副极整确知贸己环话反身

选亚么带采王策真女谈严斯况色打德告仅它气料神率识劳境源青护列兴许户马港则节款拉直案股光较河花根布线土克再群医清速律她族历非感占续师何影功负验望财类货约艺售连纪按讯史示象养获石食抓富模始住赛客越闻央席坚

份士热限米银息校均房周游千失八检足配存九命尔即防钱评复考依断范础油照段落访未额双让切须儿便空往你层低奖注黄英承远版维算破铁乐边初满病响药助致善突爱容香称购届余素请白宣健牌促培竞巴稳继紧字困刘旅声超随例

担友号显却监材且春居适除红半买充陈火搞图阳六察试太什执片古七球修尽控讲排粮武预亲挥卖审措荣洲卫希店良属险曾围域令站苏龙念罗吨器汇康减习演普田班待星飞写矿轻扩言章汽靠毛终仍景置底福止离泽波兰核降训逐票菜

座献钢眼损宁像苦印融独湖早予夫编换欧努著顾征升态套介送某斗状画留航派室临兵补宝略黑综云差纳密贫剧犯阿击遇岁阶烈督吃丰馆招害官树听庭另沙私针胜贷网愿托缺园假酒音巨既判输讨测读洋括筑欢刚庆久陆找楼激晚绝压

故互签汉草木亩短绍迎吸警藏疗贵纷授登探索湾宏录申诉秀序顺死卡歌午孩桥喜川邓扬津温船库订练候退违否彩棉帮拿罪币角召灾妇杨奋绩虽煤免笔够永圳停奥鲜朝吴岛党移尼急博贯拥束左细舞幅语俄奇般简拍脑债固威券追筹刻

映繁伟甚饭右彻烟沿街血冲洪植誉刊玉厅救潮迅伍怎付倍顿述播励斤乎纸振旧障鼓艰呼吉男绿尚夏亏季松哈祖典韩遍夜轮板抗摄杂皮贡借幕罚伤岸扶乱曲脱践危澳童散味叶累谢孙邮雄兼微呢谁惠偿署择染答块徐鱼赞课盛延瑞怀堂